Science and Technology

A BEDFORD SPOTLIGHT READER

Science and Technology

A BEDFORD SPOTLIGHT READER

Erica Duran
MiraCosta College

Lauren Mecucci Springer
Mt. San Jacinto College

For Bedford/St. Martin's
Vice President, Editorial, Macmillan Learning Humanities: Edwin Hill
Executive Program Director for English: Leasa Burton
Senior Program Manager: John E. Sullivan III
Executive Marketing Manager: Joy Fisher Williams
Director of Content Development, Humanities: Jane Knetzger
Development Editor: Lexi DeConti
Content Project Manager: Louis C. Bruno Jr.
Senior Workflow Project Manager: Lisa McDowell
Production Supervisor: Robin Besofsky
Media Project Manager: Rand Thomas
Senior Manager of Publishing Services: Andrea Cava
Project Management: Lumina Datamatics, Inc.
Composition: Lumina Datamatics, Inc.
Text Permissions Manager: Kalina Ingham
Senior Text Permissions Researcher: Elaine Kosta/Lumina Datamatics, Inc.
Photo Permissions Editor: Angela Boehler
Photo Researcher: Krystyna Borgen/Lumina Datamatics, Inc.
Director of Design, Content Management: Diana Blume
Text Design: Castle Design; Janis Owens, Books by Design, Inc.; Claire Seng-Niemoeller
Cover Design: William Boardman
Cover Image: koya79/iStock/Getty Images
Printing and Binding: King Printing Co., Inc.

Manufactured in the United States of America.

2 3 4 5 6 24 23

For information, write: Bedford/St. Martin's, 75 Arlington Street, Boston, MA 02116

ISBN 978-1-319-20718-2

Acknowledgments
Text acknowledgments and copyrights appear at the back of the book on pages 313–15, which constitute an extension of the copyright page. Art acknowledgments and copyrights appear on the same page as the art selections they cover.

About the Bedford Spotlight Reader Series

The Bedford Spotlight Reader Series is a line of single-theme readers, each featuring Bedford's trademark care and quality. The readers in the series collect thoughtfully chosen readings sufficient for an entire writing course—about thirty-five selections—to allow instructors to provide carefully developed, high-quality instruction at an affordable price. Bedford Spotlight Readers are designed to help students make inquiries from multiple perspectives, opening up topics such as borders, food, gender, happiness, monsters, science and technology, and sustainability. An editorial board of a dozen compositionists whose programs focus on specific themes have assisted in the development of the series.

Bedford Spotlight Readers offer plenty of material for a composition course while keeping the price low. Each volume in the series includes multiple perspectives on the topic and its effects on individuals and society. Chapters are built around central questions such as "What Makes People Happy?" and "Is Technology Surpassing Biology?" and so offer numerous entry points for inquiry and discussion. High-interest readings, chosen for their suitability in the classroom, represent a mix of genres and disciplines as well as a choice of accessible and challenging selections to allow instructors to tailor their approach. Each chapter thus brings to light related—even surprising—questions and ideas.

A rich editorial apparatus provides a sound pedagogical foundation. A general introduction, chapter introductions, and headnotes supply context. Following each selection, writing prompts provide avenues of inquiry tuned to different levels of engagement, from reading comprehension ("Understanding the Text"), to critical analysis ("Reflection and Response"), to the kind of integrative analysis appropriate to the research paper ("Making Connections"). An appendix, "Sentence Guides for Academic Writers," helps students with the most basic academic scenario: having to understand and respond to the ideas of others. This is a practical module that helps students develop an academic writing voice by giving them sentence guides, or templates, to follow in a variety of rhetorical situations and types of research conversations. The instructor resources tab of the catalog page for *Science and Technology* offers support for teaching, with sample syllabi, additional readings, video links, and more; visit **macmillanlearning.com/spotlight**.

Preface for Instructors

Success in higher education requires strong writing across the disciplines, yet incoming first-year students are often unaware of this, thinking that strong writing skills are only required in the humanities. The belief that the humanities and the sciences are unrelated to one another is an age-old misunderstanding, and, unfortunately, scholars in each often know little about the other and do little to bridge the gap for themselves and their students. Scientist and novelist C. P. Snow discussed this divide as far back as 1959 in his essay, "The Two Cultures," and sixty years later, the disconnect is still present.

Therefore, First-Year Composition (FYC) curriculum that evidences the interdependent relationship between Science, Technology, Engineering, and Mathematics (STEM) and other disciplines is necessary to begin mitigating the aforementioned disconnect. Students often view writing and critical thinking skills they acquire in their FYC classes as wholly disconnected from skills required of scientists and engineers. All too often STEM majors and non-STEM majors feel as if the skill sets they develop in individual classes are not useful or even applicable across all of their coursework. Admittedly, we were guilty of perpetuating this problem. As women who studied Hemingway and Dante, we didn't have any experience researching or teaching on subjects of technology or science, nor did we expect to. But we found ourselves teaching during an educational shift in which technology and science have become heavily emphasized across the curriculum. This is where *Science and Technology* can play a pivotal role in preparing students for the rigors of academic writing and critical thinking while simultaneously exposing them to the significance that technological and scientific study and advancement.

When we started teaching our FYC classes centered on a science and technology theme, we included readings from a variety of science and technology related topics. Students who were majoring in STEM found themselves able to bring in research they were doing in their science courses as part of their argumentative papers, while non-STEM majors were fascinated to learn about and evaluate scientific inventions, methods, and issues they had never heard of or previously considered in any depth. Learning composition through the lens of science and technology has allowed our students to experience writing and reading across the disciplines firsthand.

Science and Technology is a product of the curriculum we developed for our thematic FYC classes and the many rich discussions with other FYC professors that followed. The readings we've chosen are a combination of approachable and more complex science pieces, including essays, studies, and contemporary journalism. Many different voices are represented, including those of biologists, planetary scientists, generational experts, novelists, journalists, engineers, ethicists, climate scientists, and more, speaking to a variety of timely topics. The readings are organized into five chapters, representing five major areas of science and technology where students are prompted to relate the readings to their own lives. Chapter 1, "Has Technology Made Us Gods of the Natural World?," examines the power humans have over the environment and the creatures that share it, covering topics such as reviving extinct species and controlling the water crisis through large scale desalination. In Chapter 2, "Is Technology Surpassing Biology?," we turn to the field of medicine, considering issues such as the future of organ generation and the ethics of immortality. Chapter 3, "Have You Been Spied On Today?," takes a look at the prevalence of surveillance and how it's changing everything from the way we work to the way we're governed. Technology is also redefining our economy, and Chapter 4, "Who Controls the Economy, Us or the Technology We've Created?," explores drones, self-driving cars, emerging currencies, and other forces that are contributing to that change. The final chapter, Chapter 5, "How Is the Internet Defining What Matters to Society?," takes on the internet and asks readers to consider how things like smartphones, social media, and artificial intelligence are changing the way we get our news, communicate with our peers, and think about the world around us. Through these questions and more the book asks students to consider the rhetorical, philosophical, and ethical issues inherent in the developments and applications of science and technology.

Together, the selections in this book require students to write and read about current, engaging topics that span disciplines. It's designed specifically for writing courses, so a STEM background isn't needed for students to understand the readings or for instructors to facilitate a class discussion or give assignments. Each reading is introduced by a brief, informational note that gives readers the context they need to approach the essay and is followed by a set of three categories of questions: "Understanding the Text," "Reflection and Response," and "Making Connections." These questions help students to better understand the selection at hand, asking them to reinforce the concepts they've learned, and to make connections between the content and other essays in the book, and to themes in media, culture, history, and their own lives. For further connections, we include two additional tables of content organized by discipline and theme. These can be found on pages xvii–xxv.

The topics and selections in this book are meant to be accessible to any audience; we are not trying to reach the experts, just as we were not trying to reach experts while teaching our thematic classes. We were thrilled and surprised to find that when teaching these themed courses, they were fun both for our students to take and for us to teach. Often, class discussions took unexpected and organic turns that kept us engaged not only in the teaching process, but in the process of discovery along with our students, helping us to develop a passion for these topics, too. With that experience in mind, we designed this book for all composition instructors regardless of their experience with this topic.

With this interdisciplinary set of readings, our goal is to challenge both students and instructors to think about why certain sciences and technologies are accepted by society, or not, and the potential, or present, ramifications of those technologies. We want to encourage readers to think critically about how, when, and why they interact with technology. Ultimately, just as our interest in these fields grew through the experience of researching and teaching them, we hope this book is just the start of new and enriching conversations with your own composition students on the topics of science and technology.

Acknowledgments

As friends and colleagues whose lives have often paralleled one another, we are fortunate to have so many of the same people to thank for their support while we embarked on this journey to publication. At Macmillan, we first owe our gratitude to our friend, Amy O'Brien Shefferd, who not only believed in this project early on, but who was instrumental in getting it off the ground. Many thanks to another friend, Lauren Arrant, who cheered us on throughout the whole project as well. Thanks to John Sullivan, Senior Program Manager for Readers and Literature, for his excitement about this book and his expert and patient guidance. We would also like to extend our thanks to our editors, Alexandra DeConti and Cari Goldfine, for their insightful feedback throughout the editing process. We appreciate the efforts of everyone at Macmillan who brought this book to fruition, and would also like to thank Edwin Hill, Leasa Burton, Joy Fisher Williams, Louis Bruno, Kalina Ingham, Angela Boehler, Hilary Newman, and William Boardman.

This book was also strengthened deeply by the thoughtful and thorough insights of our colleagues. Thanks to Thong Nguyen and Joe Salamon from the chemistry and physics departments, respectively, at MiraCosta College, for providing feedback on our initial ideas for this book. A million thanks to our peer reviewers who contributed their feedback in the earliest stages of our proposal: Jennifer Allard, Mt. San Jacinto College; Wallace Cleaves, University of California, Riverside; Michelle Davidson, The University of Toledo; Ashley Dycus, University of West Georgia; Lynee Gaillet, Georgia State University; Karen Gardiner, University of Alabama; James Harper, University of Colorado Boulder; Jennifer Harper, University of Colorado Anschutz Medical Campus; Carrie Morrow, California State University San Marcos; Michelle Sidler, Auburn University; Ryan Sullivan, Mt. San Jacinto College; and Carrie Wastal, University of California, San Diego.

We also owe a tremendous amount of gratitude to our incredible mentors at California State University San Marcos: Catherine Cucinella, Dawn Formo, and Martha Stoddard Holmes, who have always treated us as peers long before we were; your personal and professional guidance over the years has been instrumental to our success as teachers and writers. Extra thanks to you, Catherine, whose individual feedback in the early stages of this project was invaluable. We are grateful to each of them for their continued professional and personal support, even as we have gone on to teach elsewhere. To our families, we love you. Lauren's parents watched the kids before major deadlines, and Erica's mom believed she could write a book long before she did. Finally, we are so thankful for our husbands, Andrew and Alex, for taking on even more at home while we spent countless hours holed up at the office, dining room table, and anywhere else we could find time and space to write and collaborate. We have to also thank our little ones Jackson, Jameson, Wyatt, and Annabelle who, while they aren't old enough to realize it yet, are always our constant motivation. During the writing of this book's proposal and multiple drafts, both of us simultaneously potty-trained toddlers, sent our first-born sons off to kindergarten, and worked toward or earned tenure at our respective colleges, to say nothing of the doctorate that Lauren is still in the midst of completing. As we write this acknowledgment, we are reminded of and humbled by the incredible support we have in our lives.

Erica Duran
Lauren Mecucci Springer

Bedford/St. Martin's Puts You First

From day one, our goal has been simple: to provide inspiring resources that are grounded in best practices for teaching reading and writing. For more than 35 years, Bedford/St. Martin's has partnered with the field, listening to teachers, scholars, and students about the support writers need. We are committed to helping every writing instructor make the most of our resources.

How Can We Help *You*?

- Our editors can align our resources to your outcomes through correlation and transition guides for your syllabus. Just ask us.
- Our sales representatives specialize in helping you find the right materials to support your course goals.
- Our *Bits* blog on the Bedford/St. Martin's English Community (**community.macmillan.com**) publishes fresh teaching ideas weekly. You'll also find easily downloadable professional resources and links to author webinars on our community site.

Contact your Bedford/St. Martin's sales representative or visit **macmillan learning.com** to learn more.

Print and Digital Options for *Science and Technology*

Choose the format that works best for your course, and ask about our packaging options that offer savings for students.

Print

- *Paperback*. To order the *Science and Technology* edition, use ISBN 978-1-319-20718-2.

Digital

- *Innovative digital learning space.* Bedford/St. Martin's suite of digital tools makes it easy to get everyone on the same page by putting student writers at the center. For details, visit **macmillanlearning .com/englishdigital**.
- *Popular e-book formats.* For details about our e-book partners, visit **macmillanlearning.com/ebooks**.

- *Inclusive Access.* Enable every student to receive their course materials through your LMS on the first day of class. Macmillan Learning's Inclusive Access program is the easiest, most affordable way to ensure all students have access to quality educational resources. Find out more at **macmillanlearning.com/inclusiveaccess.**

Your Course, Your Way

No two writing programs or classrooms are exactly alike. Our Curriculum Solutions team works with you to design custom options that provide the resources your students need. (Options below require enrollment minimums.)

- *ForeWords for English.* Customize any print resource to fit the focus of your course or program by choosing from a range of prepared topics, such as Sentence Guides for Academic Writers.
- *Macmillan Author Program (MAP).* Add excerpts or package acclaimed works from Macmillan's trade imprints to connect students with prominent authors and public conversations. A list of popular examples or academic themes is available upon request.
- *Bedford Select.* Build your own print handbook or anthology from a database of more than 800 selections, and add your own materials to create your ideal text. Package with any Bedford/St. Martin's text for additional savings. Visit **macmillanlearning.com/bedfordselect.**

Instructor Resources

You have a lot to do in your course. We want to make it easy for you to find the support you need—and to get it quickly. Instructor resources, including sample syllabi and a list of related resources, are available from **macmillanlearning.com.**

Contents

Contents by Discipline

Astronomy

Biology

Communications and Journalism

Composition and Literature

Computer Science

Economics

Engineering

Environmental Science

Global Studies

Government

History

Journalism

Philosophy and Ethics

Contents by Theme

Animals

Climate and the Environment

Culture and Society

Economy

Food and Agriculture

Genetics

Health and Medicine

Immortality

Internet and Social Media

Politics and Policy

Privacy and Security

Robots and Artificial Intelligence

Social Justice

Introduction for Students

How Do We Define Science and Technology?

The word *science* is used every day; its meaning is expansive, covering a myriad of topics. For example, the extensive entry on *science* in the *Oxford English Dictionary* (*OED*) dates to the fourteenth century, illustrating the term's evolution as it has grown to encompass many areas of study. For this book, science should be defined, using a portion of the *OED*'s definition, as "a branch of study that deals with a connected body of demonstrated truths or with observed facts systematically classified and more or less comprehended by general laws, and incorporating trustworthy methods (now esp. those involving the scientific method and which incorporate falsifiable hypotheses) for the discovery of new truth in its own domain." While the official definition may appear convoluted, it is simply saying that science is an agreed-upon set of truths. So how do scientists arrive at these "truths"? Ideally, it is through exhaustive observations and/or experiments — in other words, they use "trustworthy methods" to arrive at "truth." These methods are meant to ensure that no science is based solely on a singular opinion or experiment. For example, physics, the study of interactions between matter and energy and the laws that govern the physical world, is one type of science.

Just as the definition of science is exhaustive, so, too, is the way we define technology. Again from the *OED*, technology is defined as "the application of [scientific] knowledge for practical purposes, esp. in industry, manufacturing, etc." In other words, science is the study, and technology is the practice. If we think again about physics and now imagine how that science is applied in the real world, it is difficult to name a type of technology that does not rely on physics. GPS navigation systems are just one example of a technology that would not exist without physics — Albert Einstein's theory of relativity to be exact. So while science and technology have their own definitions, the two studies often overlap, creating a complex and symbiotic relationships between the two subjects.

Why Study Science and Technology?

"We live in a society exquisitely dependent on science and technology, in which hardly anyone knows anything about science and technology. This is a clear prescription for disaster," argued astronomer Carl Sagan in his 1990 article "Why We Need to Understand Science." He argues that our superficial knowledge is dangerous, in part, because scientific discovery is critical to human success. Thirty years later, Sagan's observation still holds true – possibly even more true: today's technology is more complex, more embedded, and more relied upon than ever before.

Let's think about this for a moment – what is a piece of technology you use every day? Do you know all of its capabilities beyond the few tasks you use it for? For example, think about a smartphone. For the most part, you probably use your phone to access your social media accounts, send texts, and make the occasional phone call. This may not seem like much, especially in light of all that the smartphone is capable of, but have you spent any time thinking about *how* it refreshes your Instagram feed in mere seconds? In the last thirty years, cellular phones have morphed from flip phones that were only able to make and receive calls, to today's smartphones, which are essentially pocket-sized computers, performing many of the duties that used to be the sole purview of desktops, sometimes more efficiently. A brief Google search will probably reveal smartphone capabilities that you were unaware of, and that is exactly Sagan's point.

While many of us may not think deeply about scientific and technological progress, we all feel the impact of science and technology on a daily basis. In recent years American society has placed increased emphasis on learning about science, technology, engineering, and math (commonly known as STEM) at all levels of education. In "Most Americans Say U.S. STEM Education is Middling, New Poll Finds," author Caroline Preston provides a brief history of why and when the United States became so invested in STEM education. In 2006, President George W. Bush announced the "American Competitiveness Initiative" in an effort to bolster STEM education with the clear goal of advancing innovation. Three years later, in 2009,

President Barack Obama began a program to train 100,000 new STEM teachers, with the specific goal of attracting more girls and minorities to the STEM field. However, despite efforts over the past few decades, Americans still score lower today in math and the sciences than students from many other developed countries. Sagan's concern is still relevant, and as science and technology continue to advance, America's lag is only set to grow. Therefore, as students, teachers, readers, and human beings who engage with science and technology every day, it is essential that we understand science and technology's significant effects on us, so that we can ensure that these advances are used wisely and safely. In fact, a deeper understanding is essential not just for those of you who will go on to pursue careers in the fields of science and tech: this is necessary for each of us to be informed citizens in today's world.

As part of an ever more tech-dependent society, you do not need to be a scientist to study, analyze, or even care about science. The interdependent relationship between our lives and science and technology requires us to consider not only how the technologies we interact with and depend on function but also to evaluate the decisions surrounding these technologies. Let's think again about the smartphone — most people who own a smartphone would agree they "need" it. But at what point did users begin to accept the risks that are associated with their phones in order to meet that "need"? Stuart Sumner's excerpted *You: For Sale* (p. 160) examines the user agreements people encounter when they download, for example, a new, seemingly harmless app that in fact leaks your personal information; Sumner describes how the government has taken advantage of these user agreements, collecting data on people simply because people clicked "I Agree" to the terms of use. These user agreements, while often legal and necessary, do raise some ethical questions about why we give technology complete access to our lives and personal information.

The "permission" users grant to apps or websites often only requires the click of a button, and the vast majority of users admit to never reading the fine print. In a 2016 study, Jonathan Obar of York University in Toronto and Anne Oeldorf-Hirsch of the University of Connecticut, both professors of communication, tested the theory that users don't read, or at best only skim, the fine print. Because their study was designed to test how well users understood the terms they were agreeing to, the professors included a clause that required users to relinquish their first-born child to the social network they were agreeing to access. Fortunately, the social network and the under-agreement were fake: of the 543 users who were given the fake user agreement, all 543 agreed to its terms, seemingly unaware of the extreme requirement.

Why does this matter? If users are blatantly cavalier about user agreements for apps and websites, are they any more careful about investigating other technologies before using them? As a society, we tend to assume that if something is made for public consumption, then it is generally safe. After all, medical treatments are tested prior to being prescribed, and cars are tested prior to being sold. We repeatedly rely on others to develop technologies that are safe for us to use. Yet, when something turns out not to be what we expected or wanted, we rarely admit we did little to understand it in the first place, again, evidencing Sagan's concern. So, how do we begin to mitigate this growing problem? How do we educate ourselves about topics that we may or may not be otherwise studying?

An Interdisciplinary Approach

The study of science and technology, just like the study of the humanities, should be an interdisciplinary inquiry. Where empirical research is the main tenet of the natural sciences, the humanities seek to examine human culture. The number of intersections between the sciences and human culture are endless; therefore, one goal of this book is to look at science through a variety of lenses, asking questions about the motivations humans have when they create and apply science and technology.

Let's look at how interconnected the world has become as a result of the internet as just one example. In his book *You Can Do Anything: The Surprising Power of a "Useless" Liberal Arts Education* (2017), author and journalist George Anders reminds us that "the more we automate the routine stuff, the more we create a constant low-level hum of digital connectivity, [...] the more essential it is to bring human judgment into the junctions of our digital lives." Anders takes Sagan's idea and pushes it a step further, arguing the more normalized science and technology become for us, the more vital it becomes we stop, assess it, and judge it. For instance, many of us would agree that data is a powerful impetus for change, but data requires interpretation. Where did the data come from? Who gathered it and for what purpose? Were there blind spots in the gathering process? When data is gathered under controversial circumstances, as was the case during World War II when Nazis experimented on concentration camp prisoners, should that data be used or discarded? Who makes that decision? Should a doctor be able to take biological samples during a medical procedure, as was the case for Henrietta Lacks, whose cancer cells were immortalized without her consent, and then use that sample to create a groundbreaking medical treatment? Who gets the credit for that success? The patient who may have unknowingly provided that material or the scientist who made the breakthrough? All of these questions help us more fully understand the human context behind the science and technology that we rely on. Studying STEM through a humanities lens is essential because science and technology are, at their core, discovered and carried out by people.

How Is This Book Organized?

Each chapter begins by summarizing the specific readings for that particular topic; the chapter introductions are important: they help situate you in the context you will delve into. Readings are preceded by a brief introduction that offers background on the author(s) and subject of the selection. The introductions are designed to help you think about the topic before you even begin reading. Each reading, then, is followed by three sets of questions. The questions in the apparatus, especially in the "Making Connections" section, encourage you to engage in meaningful and current research outside of the text about issues related to the reading; this will help you generate your own interest, finding the most current information on the topic, and discussing any changes which have most certainly occurred following the publication of our readings. This book is organized around five topics – each framed as a question – that encourage readers to think deeply about a specific sphere where science and technology affect our daily lives.

The first chapter, "Has Technology Made Us the Gods of the Natural World?," investigates humanity's various impacts on the environment, natural resources, and other species. The readings in this chapter explore the roles technology might play in addressing some of the problems we've created for ourselves, as well as the differing motivations behind addressing these problems.

The second chapter, "Is Technology Surpassing Biology?," challenges readers to think about the innumerable ways in which technology already supplements, enhances, and transforms the human body, as well as the motivations driving newer and "better" developments. The readings look at everything from the powerful and controversial gene-editing technology known as CRISPR, to the long-standing human desire to obtain immortality.

In Chapter 3, "Have You Been Spied On Today?," we explore and examine the ways technology is allowing companies and governments to spy on groups of people and even individuals. The chapter's selections look at how everyday technologies (e.g. cell phones) and more complex technologies (e.g. meta data collected on the internet) allow other people inside our lives, often without either party's knowledge.

"Who Controls the Economy, Us or the Technology We've Created?," the fourth chapter, focuses on the ways technological advances are affecting, both negatively and positively, the economy. As technology advances and evolves, so, too, does the job market. The selections in Chapter 4 suggest that while new technologies are often fun and helpful, they may also require society to make unexpected adjustments.

The fifth and last chapter in this book, "How Is the Internet Defining What Matters to Society?," focuses on the internet's role in shaping and even reshaping societal values. From social media–powered activism to the ubiquity of fake news, this chapter's readings encourage readers to consider whether internet access is now a prerequisite for being an effective member of any community, online or otherwise.

A Note about the Readings

We recognize that the texts we have chosen, and the questions we are asking, represent a moment in time; in fact, the sciences presented here may well have evolved, or in some cases not developed further, since publication. For example, in the article "The CRISPR Pioneers" (p. 82), the authors discuss a groundbreaking and constantly developing method of gene editing that is now being applied in a myriad of ways, leaving open the possibility for you to investigate the ongoing effects of this technique. How is it being applied now? What new benefits or concerns have been uncovered as this technology has become more widely used? In "Reversing Extinction" (p. 14), the author discusses the various animals that scientists are looking to bring back to life. Did they manage to do so since the book was written? Have more animals been added to the list? Taken off the list? Why? Additionally, these readings, and their apparatuses, allow you to engage the texts in meaningful ways while also directing you to investigate new and emerging scientific methods that, even during our research process, were being used in new fields. Readers should consider the value of these topics at the time they were published along with their current relevance. The book's nonfiction readings are a mixture of scholarly articles, dealing with issues that are likely to continue to be relevant in the fields they discuss, as well as readings from more popular sources like *Scientific American* or *Time* that discuss more recent developments in specific fields.

The readings we've chosen for this book reflect our desire to make these topics as accessible and relatable as possible for readers across the disciplines. To that end, you'll notice that many of the readings here are on scientific topics written for a general audience rather than scientific papers and empirical studies. While scientific and empirical studies have significant value, they also have a very different audience than this book is designed for. Scientific papers are typically published in professional journals and read by other experts in that or similar fields, often serving as a challenging read even to those with a scientific background. For you, a general reader who may not have experience in the sciences, the readings we've chosen provide a more manageable introduction to each of these topics, encouraging you to dig deeper into the primary resources you choose to further examine.

In addition to readings by scientists and experts in their fields, many of these readings are written by nonscientists, which provide readers with an opportunity to question exactly how scientific information is typically transmitted to the general public. What is a journalist's responsibility to the reader? How does a journalist credibly and effectively write for a general audience about these complex concepts and methods, especially when he or she is also learning about concepts in order to write about them? All of this is especially important to consider for those of us who are not scientists and aren't necessarily able to immediately determine whether a concept is presented accurately.

While the majority of these readings are nonfiction, we have included brief selections of fiction throughout the book as well, so that readers from all disciplines and varied backgrounds can critically think about the ways science and technology are presented in another genre, specifically the humanities. Fiction, specifically science fiction like Kurt Vonnegut's short story collected here, "Harrison Bergeron" (p. 180), often forces readers to examine real-world issues in a creative, more approachable way, engendering readers' ability to critically think about real technological issues. "Harrison Bergeron" encourages the reader to consider how technology could be misused by a government with the desire to spy on its citizens. Because we live in a world where we routinely invite technology into our homes that could be used against us (webcams have already earned this negative reputation), this short story provokes critical questioning about where our society might be headed. Although "Harrison Bergeron" is hyperbolic, it forces readers to question whether this dystopian future is possible based on technology that exists today.

Finally, in the process of developing this book, we have had to make research and rhetorical decisions as authors; you will need to make similar decisions as you write your own projects. We could easily write a much longer book, but we had limitations, so we have chosen to omit pieces we love in order to keep others we feel are more timely, useful, or significant. As writers, you, too, may find a topic you love and could say so much more about, but find yourself having to stop researching and just finish writing! Furthermore, you may find that these readings lead you to make connections to topics not covered in this book, and we hope you will explore those opportunities to conduct your own research, too.

In summary, our goal with this book is to help promote a deeper interest in science and technology for all readers, specifically how it affects the world we all live in. But interest alone is not enough. We also hope you will begin to think critically and deeply about these fields, perhaps even developing both a greater appreciation for and skepticism of them. Although skepticism is often portrayed negatively, Sagan argues, in his 1997 book *The Demon-Haunted World*, that it is necessary for an informed society: "What skeptical thinking boils down to is the means to construct, and to understand, a reasoned argument and — especially important — to recognize a fallacious or fraudulent argument. The question is not whether we like the conclusion that emerges out of a train of reasoning, but whether the conclusion follows from the premise or starting point and whether that premise is true." In other words, let evidence guide you. Rather than passively accepting that science and technology are created by others, by experts, for reasons beyond our understanding, we can become engaged participants in the conversations that lead to their development and use. If science and technology are used by all, then all should actively discuss and question them.

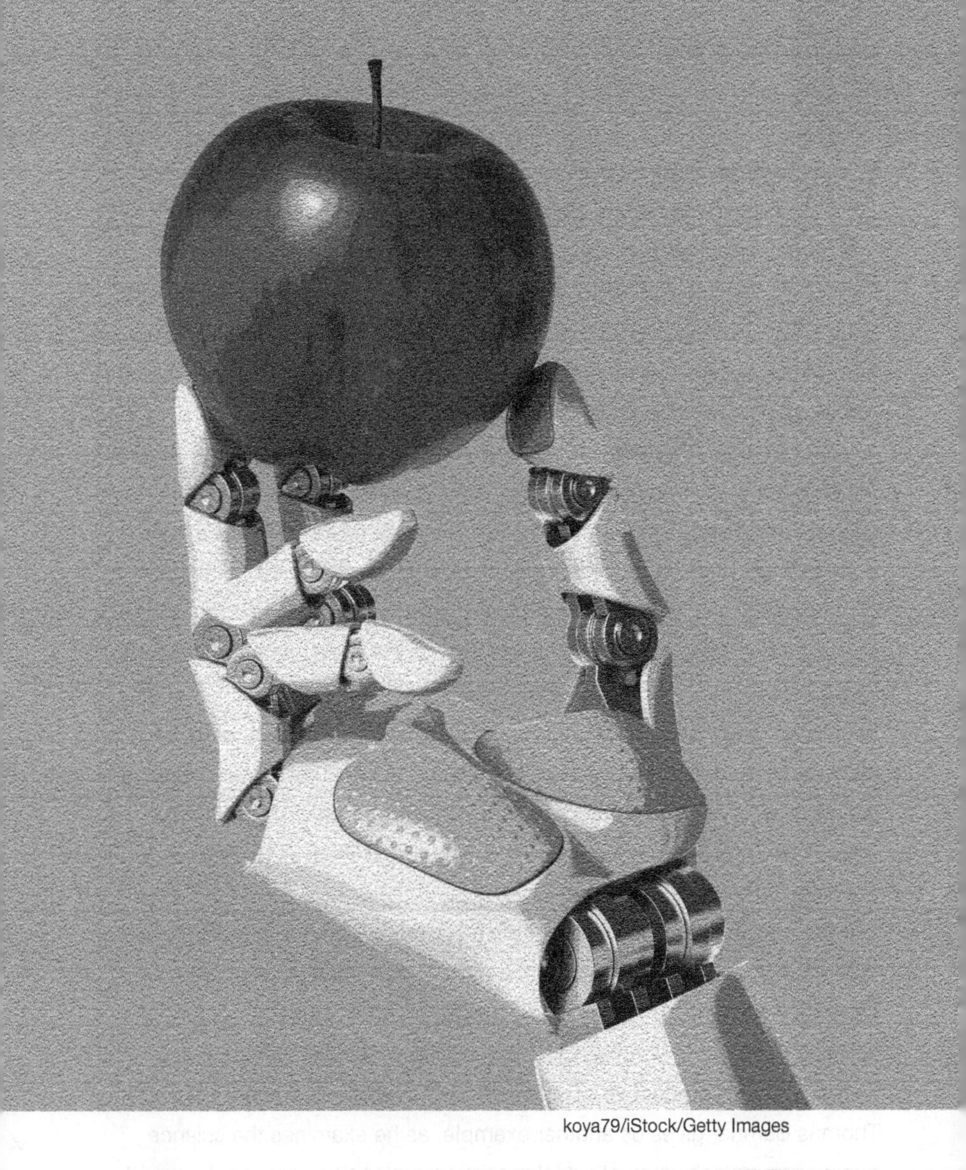

koya79/iStock/Getty Images

1 Has Technology Made Us the Gods of the Natural World?

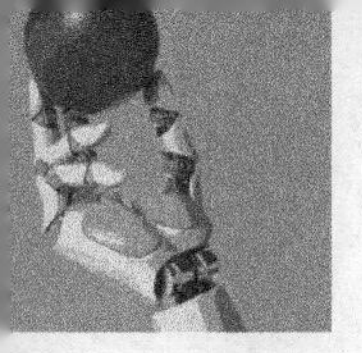

We humans have always manipulated our environment, and even the natural biology of living things — consider dog breeding, for example — but now, perhaps more than ever before, we can use science and technology to reimagine and even redesign nature. The readings in this chapter present the myriad of reasons, ranging from dire necessity to simple curiosity, that scientists and society as a whole are motivated to undertake such modifications.

Beth Shapiro's "Reversing Extinction," which opens this chapter, tackles the ever-growing need to save endangered species and goes so far as to suggest using technology to bring back some species that have already gone extinct — think about all of the rhinos we have poached to the brink of extinction. Describing a quite different way to employ technology in the animal world, Emily Anthes's "Animals Bow to Their Mechanical Overlords" introduces us to biomimetic robots that scientists are using in an attempt to help living animal and insect communities thrive. While the reasons for human intervention vary greatly depending on the circumstance of each species, both readings encourage us to learn more about the countless ways humans are reshaping the animal kingdom. However, human intervention doesn't stop there.

Humans "manage" their environment by redrawing, redistributing, and redirecting land, water, and other natural resources to fit our needs, especially when those resources are scarce or inconveniently located. Building dams, for example, is just one way in which we reshape our environment for productive compatibility. In "Quenching Society's Thirst," Thomas Sumner gives us another example, as he examines the science behind turning saltwater into drinking water; major cities around the world could immediately benefit from this technology, because they are about to run dry — a very real problem that Cape Town, South Africa, narrowly avoided, or perhaps only delayed, in 2018. To address the lack of fresh produce in urban areas, and the increasing costs of running traditional

photo: koya79/iStock/Getty Images

farms, some entrepreneurs are turning to vertical farming, a concept Selina Wang introduces us to in "The Future of Farming Is Looking Up." Clive Hamilton's "Why Geoengineering?" explores some ways that geoengineers are tackling climate change; the industry is working on a variety of solutions, some far-fetched, others more realistic, to address the consequences of a warming planet. The chapter's final two readings ask us to consider our future on this planet, and beyond. "Here's How Much It Would Cost to Travel to Mars," by Rob Wile and Pascal Lee, approaches the more fanciful idea of exploring and even colonizing another planet in case Earth becomes unsustainable, and Francesc Torralba, in "The Argument of Future Generations," encourages us all to consider what ethical responsibility we have beyond our own lifetimes.

One could easily argue that some of the ideas presented in this chapter and elsewhere in this book sound fantastical; as you read, consider the intrinsic value of these ideas. Yes, humans have the ability to control our environment to some degree, for better and for worse. But the more we dabble in the construction of the environment and interfere with the species that live in it, we must remember to constantly ask ourselves why we are doing so. Who benefits? Who loses? And, who is it that ultimately gets to decide whether we should take any of these actions? As you read through this chapter, think about the many ways you are already affected by humankind's management of the natural world, and how all of us might be affected by any one of these new technologies.

Reversing Extinction; from *How to Clone a Mammoth: The Science of De-Extinction*

Beth Shapiro

Beth Shapiro, an ecology and evolutionary biologist at the Paleogenomic Lab at the University of California, Santa Cruz, has spent her career examining the DNA of ancient species in an effort to better understand their evolution and connections to the ecosystem. She has coauthored dozens of academic articles, one book, and gave a *TED talk* on the science of de-extinction. In the following excerpt from her book *How to Clone a Mammoth: The Science of De-Extinction* (2009), Shapiro describes the actual science being developed to bring back extinct species, and she examines some of the ethical questions that arise from developing this science.

If this science sounds familiar, it is because we've all seen it play out through the fiction of Michael Crichton's *Jurassic Park* and subsequent "worlds." Although Crichton's plotline wasn't even a remote possibility in 1990, when his book debuted, the story reignited an age-old interest in humanity's power over nature through the use of science and technology, and the story specifically fueled a fascination with bringing back extinct species. The science has advanced since Crichton's book first appeared, but Shapiro's writing reminds us that if scientists put their minds to something, the fantastical can sometimes become reality. But, should it?

The Sixth Extinction

More than 3,700 years after the last mammoth died on Wrangel Island, we are witnessing an alarming number of contemporary extinctions, and the rate of extinction appears to be increasing. Some scientists have gone so far as to refer to the Holocene extinctions as the Sixth Extinction, suggesting that the crisis in the present day has the potential to be as destructive to Earth's biodiversity as the other five mass extinctions in our planet's history.

The word alone—extinction—frightens and intimidates us. But why should it? Extinction is part of life. It is the natural consequence of speciation and evolution. Species arise and then compete with each other for space and resources. Those that win survive. Those that lose go extinct. More than 99 percent of species that have ever lived are now extinct. Indeed, our own species' dominance is possible only because the extinction of the dinosaurs made space for mammals to diversify, and eventually we outcompeted the Neandertals.

I think people are scared of extinction for three reasons. First, we fear missed opportunities. A species that is lost is gone forever. What if that species harbored a cure for some terrible disease or was critically important in keeping our oceans clean? Once that species is gone, so is that opportunity. Second, we fear change. Extinction changes the world around us in ways that we both can and cannot anticipate. Every generation thinks of our version of the world as the authentic version of the world. Extinction makes it harder for us to recognize and feel grounded in the world we know. Third, we fear failure. We enjoy living in a rich and diverse world and feel an obligation, as the most powerful species that has ever lived on this planet, to protect this diversity from our own destructive tendencies. Yet we chop down forests and destroy habitats. We hunt and poach species even when we know they are perilously close to extinction. We build cities, highways, and dams and block migration routes between populations. We pollute the oceans, rivers, land, and air. We move around as fast as we can on airplanes, trains, and boats and introduce foreign species into previously undisturbed habitats. We fail to live up to our obligation to protect or even coexist with the other species with which we share this planet. And when we stop to think about it, it makes us feel terrible.

Extinction is much easier for us to swallow when it is clearly *not* our fault. Why did the mammoth go extinct? As humans, we want the answer to be *something natural*. Natural climate change, for example. We would

If the woolly mammoth were to be brought back from extinction, would we return it to a natural habitat? Could that even exist without human interaction?

RAUL MARTIN/National Geographic Image Collection

prefer to learn that mammoths went extinct because they needed the grasslands of the steppe tundra to survive and that they simply starved to death as the steppe tundra disappeared after the last ice age. We would prefer not to learn that mammoths went extinct because our ancestors greedily harvested them for their meat, skins, and fur.

While some of us may not care about extinction as long as we are not personally affected, many of us find extinction unacceptable, particularly if it is our fault. Most contemporary extinctions are easy to ignore, as they have little influence on our day-to-day lives. The cumulative effect of these extinctions is, however, a future of very reduced biodiversity. This future could be one in which so many changes have occurred to the terrestrial and marine ecosystems that we, ourselves, are suddenly vulnerable to extinction. It doesn't get much more personal than that.

5

Reversing Extinction

It's not completely surprising that the idea of *de*-extinction—that we might be able to bring species that have gone extinct back to life—has attracted so much attention. If extinction is not forever, then it lets us off the hook. If we can bring species that we have driven to extinction back to life, then we can right our wrongs before it is too late. We can have a second chance, clean up our act, and restore a healthy and diverse future, before it is too late to save our own species.

While it is still not possible to bring extinct species back to life, science is making progress in this direction. In 2009, a team of Spanish and French scientists announced that a clone of an extinct Pyrenean ibex, also known as a bucardo, was born in 2003 to a mother who was a hybrid of a domestic goat and a different species of ibex. To clone the bucardo, the scientists used the same technology that had been used in 1996 to successfully clone Dolly the sheep. That technology requires living cells, so in April 1999, ten months before her death, scientists captured the last living bucardo and took a small amount of tissue from her ear. They used this tissue to create bucardo embryos. Only one of 208 embryos that were implanted into the surrogate mothers survived to be born. Unfortunately, the baby bucardo had major lung deformity and suffocated within minutes.

In 2013, Australian scientists announced that they successfully made embryos of an extinct frog—the Lazarus frog—by injecting nuclei from Lazarus frog cells that had been stored in a freezer for forty years into a donor cell from a different frog species. None of the Lazarus frog embryos survived for more than a few days, but genetic tests confirmed that these embryos did contain DNA from the extinct frog.

The Lazarus frog and bucardo projects are only two of the several de-extinction projects that are under way today. These two projects involve using frozen material that was collected prior to extinction and, consequently, are among the most promising of the existing de-extinction projects. Other de-extinction projects, including mammoth and passenger pigeon de-extinction, face more daunting challenges, of which finding well-preserved material is only one. These projects are proceeding nonetheless and, in the case of the mammoth, along several different trajectories. Akira Iritani of Japan's Kinki University is trying to clone a mammoth using frozen cells and claims that he will do so by 2016. George Church at Harvard University's Wyss Institute is working to bring the mammoth back by engineering mammoth genes into elephants. Sergey Zimov of the Russian Academy of Science's Northeast Science Station worries less about about how mammoths will be brought back than about what to do with them when it happens. He established Pleistocene Park near his home in Siberia and is preparing his park for the impending arrival of resurrected mammoths.

Not all de-extinction projects take a species-centric view. George Church's project is focusing on reviving mammoth-like traits in elephants, for example. While the goal of this project is to create an animal that is mammoth-like, its motivation is to reintroduce elephants into the Arctic. Stewart Brand and Ryan Phelan have taken an even more holistic view. Together, they created a nonprofit organization called Revive & Restore, and are asking people to consider all the ways in which de-extinction and the technology behind it might change the world over the next few decades or centuries. In addition to initiating the passenger pigeon de-extinction project, Revive & Restore is driving several projects to revive living species that have dangerously low amounts of genetic diversity. With Oliver Ryder of San Diego's Frozen Zoo, for example, Revive & Restore is isolating DNA from archived remains of black-footed ferrets, which are nearly extinct in the present day. They hope to identify genetic diversity that was present in black-footed ferrets prior to their recent decline and, using de-extinction technologies, to engineer this lost diversity back into living populations. 10

In March 2013, Revive & Restore organized a TEDx event at National Geographic's headquarters in Washington, DC, to focus on the science and ethics of de-extinction. This media event was the first attempt to address de-extinction at a more sophisticated level than attention-grabbing headlines. When the event concluded, public opinion about de-extinction was mixed. Some people loved and others hated the idea that extinctions might be reversed. Fears were expressed about the uncertain environmental impacts of reintroduced resurrected species. Some ethicists argued that de-extinction is morally wrong; others insisted that it is morally wrong *not* to bring things back to life, if indeed it were possible to do so. Voices were also raised in opposition to the cost of de-extinction and whether the potential benefits justified this cost. What was lost in the noise of the ensuing public debates, however, was discussion of the current state of the science of de-extinction: what is possible now, and what will ever be possible? And, perhaps more importantly, there was little conversation and certainly no consensus about what the goal of de-extinction should be. Should we focus on bringing species back to life or on resurrecting extinct ecosystems? Or should the focus be on preserving or invigorating ecosystems in the present day? Also, and importantly, what constitutes a successful de-extinction?

"The technology to do all of this is available today. But what would the end product of this experiment be?"

In this book, I aim to separate the science of de-extinction from the science fiction of de-extinction. I will describe what we can and cannot do today and how we might bridge the gap between the two. I will argue that the present focus on bringing back particular species—whether that means mammoths, dodos, passenger pigeons, or anything else—is misguided. In my mind, de-extinction has a place in our scientific future, but not as an antidote to extinctions that have already occurred. Extinct species are gone forever. We will never bring something back that is 100 percent identical—physiologically, genetically, and behaviorally identical—to a species that is no longer alive. We can, however, resurrect some of their extinct traits. By engineering these extinct traits into living organisms, we can help living species adapt to a changing environment.

We can reestablish interactions between species that were lost when one species went extinct. In doing so, we can revive and restore vulnerable ecosystems. This—the resurrection of ecological interactions—is, in my mind, the real value of de-extinction technology.

A Scientific View of De-Extinction

I am a biologist. I teach classes and run a research laboratory at the University of California, Santa Cruz. My lab specializes in a field of biology called "ancient DNA." We and other scientists working in this field develop tools to isolate DNA sequences from bones, teeth, hair, seeds, and other tissues of organisms that used to be alive and use these DNA sequences to study ancient populations and communities. The DNA that we extract from these remains is largely in terrible condition, which is not surprising given that it can be as old as 700,000 years.

During my career in ancient DNA, I have extracted and studied DNA from an assortment of extinct animals including dodos, giant bears, steppe bison, North American camels, and saber-toothed cats. By extracting and piecing together the DNA sequences that make up these animals' genomes, we can learn nearly everything about the evolutionary history of each individual animal: how and when the species to which it belonged first evolved, how the population in which it lived fared as the climate changed during the ice ages, and how the physical appearance and behaviors that defined it were shaped by the environment in which it lived. I am fascinated and often amazed by what we can learn about the past simply by grinding up and extracting DNA from a piece of bone. However, regardless of how excited I feel about our latest results, the most common question I am asked about them is, "Does this mean that we can clone a mammoth?" 15

Always the mammoth.

The problem with this question is that it assumes that, because we can learn the DNA sequence of an extinct species, we can use that sequence to create an identical clone. Unfortunately, this is far from true. We will never create an identical clone of a mammoth. Cloning, as I will describe later, is a specific scientific technique that requires a preserved *living* cell, and this is something that, for mammoths, will never be found.

Fortunately, we don't have to clone a mammoth to resurrect mammoth traits or behaviors, and it is in these other technologies that de-extinction research is progressing most rapidly. We could, for example, learn the DNA sequence that codes for mammoth-like hairiness and then change the genome sequence of a living elephant to make a hairier elephant. Resurrecting a mammoth trait is, of course, not the same thing as resurrecting a mammoth. It is, however, a step in that direction.

Scientists know much more today than was known even a decade ago about how to sequence the genomes of extinct species, how to manipulate cells in laboratory settings, and how to engineer the genomes of living species. The combination of these three technologies paves the way for the most likely scenario of de-extinction, or at least the first phase of de-extinction: the creation of a healthy, living individual.

First, we find a well-preserved bone from which we can sequence the complete genome of an extinct species, such as a woolly mammoth. Then, we study that genome sequence, comparing it to the genomes of living evolutionary relatives. The mammoth's closest living relative is the Asian elephant, so that is where we will start. We identify differences between the elephant genome sequence and the mammoth genome sequence, and we design experiments to tweak the elephant genome, changing a few of the DNA bases at a time, until the genome looks a lot 20 more mammoth-like than elephant-like. Then, we take a cell that contains one of these tweaked, mammoth-like genomes and allow that cell to develop into an embryo. Finally, we implant this embryo into a female elephant, and, about two years later, an elephant mom gives birth to a baby mammoth.

The technology to do all of this is available today. But what would the end product of this experiment be? Is making an elephant whose genome contains a few parts mammoth the same thing as making a mammoth? A mammoth is more than a simple string of As, Cs, Gs, and Ts—the letters that represent the nucleotide bases that make up a DNA sequence. Today, we don't fully understand the complexities of the transition from simply stringing those letters together in the correct order—the DNA sequence, or genotype—to making an organism that looks and acts like the living thing. Generating something that looks and acts like an extinct species will be a critical step toward successful de-extinction. It will, however, involve much more than merely finding a well-preserved bone and using that bone to sequence a genome.

When I imagine a successful de-extinction, I don't imagine an Asian elephant giving birth in captivity to a slightly hairier elephant under the close scrutiny of veterinarians and excited (and quite possibly mad) scientists. I don't imagine the spectacle of this exotic creature in a zoo enclosure, on display for the gawking eyes of children who'd doubtless prefer to see a *T. rex* or *Archaeopteryx* anyway. What I do imagine is the perfect arctic scene, where mammoth (or mammoth-like) families graze the steppe tundra, sharing the frozen landscape with herds of bison, horses, and reindeer—a landscape in which mammoths are free to roam, rut, and reproduce without the need of human intervention and without fear of re-extinction. This—building on the successful creation of one individual to produce and eventually release entire populations into the wild—constitutes the second phase of de-extinction. In my mind, de-extinction cannot be successful without this second phase.

The idyllic arctic scene described above might be in our future. However, before a successful de-extinction can occur, science has some catching up with the movies to do. We have yet to learn the full genome sequence of a mammoth, for example, and we are far from understanding precisely which bits of the mammoth genome sequence are important to make a mammoth look and act like a mammoth. This makes it hard to know where to begin and nearly impossible to guess how much work might be in store for us.

Another yet-to-be-solved problem is that some important differences between species or individuals, such as when or for how long a particular gene is turned on during development or how much of a particular protein is made in the gut versus in the brain, are inherited epigenetically. That means that the instructions for these differences are not coded into the DNA sequence itself but are determined by the environment in which the animal lives. What if that environment is a captive breeding facility? Baby mammoths, like baby elephants, ate their mother's feces to establish a microbial community capable of breaking down the food they consumed. Will it be necessary to reconstruct mammoth gut microbes? A baby mammoth will also need a place to live, a social group to teach it how to live, and, eventually, a large, open space where it can roam freely but also be safe from poaching and other dangers. This will likely require a new form of international cooperation and coordination. Many of these steps encroach on legal and ethical arenas that have yet to be fully and adequately defined, much less explored.

Despite this somewhat pessimistic outlook, my goal for this book is not to argue that de-extinction will not and should never happen. In fact, I'm nearly certain that someone will claim to have achieved de-extinction within the next several years. I will argue, however, for a high standard by which to accept this claim. Should de-extinction be declared a success if a single mammoth gene is inserted into a developing elephant embryo and that developing elephant survives to become an adult elephant? De-extinction purists may say no, but I would want to know how inserting that mammoth DNA changed the elephant. Should de-extinction be declared a success if a somewhat hirsute elephant is born with a cold-temperature tolerance that exceeds that of every living elephant? What if that elephant not only looks more like a mammoth but is also capable of reproducing and sustaining a population where mammoths once lived? While others will undoubtedly have different thresholds for declaring de-extinction a success than I do, I argue that this—the birth of an animal that is capable, thanks to resurrected mammoth DNA, of living where a mammoth once lived and acting, within that environment, like a mammoth would have acted—is a successful de-extinction, even if the genome of this animal is decidedly more elephant-like than mammoth-like.

Making De-Extinction Happen

Many technical hurdles stand in the way of de-extinction. While science will eventually find a way over these hurdles, doing so will require significant investment in both time and capital. De-extinction will be expensive. There will be important issues to consider about animal welfare and environmental ethics. As with any other research project, the cost to society of the research needs to be weighed against the gains to society of what might be learned or achieved.

If we brought back a mammoth and stuck it in a zoo, then we could study how mammoths are different from living elephants and possibly learn something about how animals evolve to become adapted to cold climates. Some scientists who favor de-extinction see this as a reasonable goal, and many nonscientists would be just as happy to see unextinct species in zoos as they would be to see them in safari parks or unmanaged wild habitat. But is bringing a mammoth back to life so that we can look at it and possibly study it enough of a societal gain to justify the costs of creating that mammoth?

If, like elephants, mammoths helped to maintain their own habitat, then bringing mammoths back to life and releasing them into the Arctic may transform the existing tundra into something similar to the steppe tundra of the ice ages. This might create habitat for living and endangered Arctic species, such as wild horses and saiga antelopes, and other extinct megafauna that might be targets for de-extinction, such as short-faced bears. Is the possibility of revitalizing modern habitats in a way that benefits living species enough to justify the expense? Of course, ecosystems change and adapt over time, and there is no certainty that the modern tundra would convert back to the steppe tundra of the Pleistocene even with free-living populations of unextinct mammoths. Should uncertainty of success influence our analysis of the cost of de-extinction?

What if we identified a very recently extinct species that played a similarly important role in a present-day environment, and brought that species back to life? For example, kangaroo rats are native to the deserts of the American Southwest, but their populations have become increasingly fragmented over the last fifty years, and many subspecies are known to be extinct today. Kangaroo rats are so important to their ecosystem that their disappearance can cause a desert plain to turn into arid grassland in less than a decade. The domino effects of kangaroo rat extinction include the disappearance of plants with small seeds and their replacement by plants with larger seeds (on which the kangaroo rat would have fed), in turn leading to a decline in seed-eating birds. The decrease in foraging and burrowing slows plant decomposition and snowmelt, and the lack of burrows leaves many smaller animal and insect species without shelter. When the kangaroo rat goes extinct, the entire ecosystem is in danger of the same. If bringing the kangaroo rat back could save the entire ecosystem, would that be sufficient to justify the expense?

In the chapters that follow, I will walk through the steps of de-extinction. As I indicated earlier, de-extinction is likely to happen in two phases. The first phase includes everything up to the birth of a living organism, and the second will involve the production, rearing, release, and, ultimately, management of populations in the wild. For each step

30

in the process, I will describe what we now know, what we need to know, what we are likely to know soon, and what's likely to remain unknown. I will discuss both the science and the ethical and legal considerations that are likely to be part of any de-extinction project. Although the book is organized as a how-to manual, de-extinction is not a strictly linear process, and not all steps will apply to every species. Species from which living tissue was cryopreserved prior to their extinction may be clonable in the traditional sense, for example, while other species will require additional steps to create a viable embryo.

As part of my professional relationship with Revive & Restore, I have been involved in research that focuses on two species—mammoths and passenger pigeons—that are presently targets of de-extinction efforts. This will no doubt result in an animal-centric (really mammoth and pigeon-centric) view of the process. Still, many of the details will be broadly applicable across taxonomic lines. My hope is to present a realistic but not cynical view of the prospects for de-extinction, which I believe has the potential to be a powerful new tool in biodiversity conservation.

Understanding the Text

1. Shapiro explains that scientists will never be able to actually clone an extinct animal – only "resurrect some of [its] extinct traits" (par. 12). Why is bringing back an extinct animal, in the way it originally lived and looked, impossible?
2. In describing George Church's plans to bring back the mammoth, Shapiro states that "[n]ot all de-extinction projects take a species-centric view" (par. 10). What does she mean by this? How might Church's project be an example of this?

Reflection and Response

3. What credentials does Shapiro list, and do they make her a credible source? Based on information she provides in the text, what do you know about her as a scientist and as a writer? What is implied? What more do you want to know about her?
4. Think about the tone of this piece. Who do you believe is Shapiro's intended audience, and what is the purpose (stated or implied) of her book? Based on this excerpt, does she appear to be effective in accomplishing this purpose?
5. This article has obvious ties to *Jurassic Park* and *Jurassic World* and the concerns constantly expressed throughout the franchise of books and movies. What connections do you see?

Making Connections

6. Shapiro mentions the arctic several times and draws attention to the fact that it is not a stable environment in which to re-introduce a species.

Why not? Research some of the current environmental challenges the arctic is facing and the consequences those are having for native species. Based on your research, do you think a woolly mammoth hybrid would be successful if returned to this environment? Why or why not? How might the state of its habitat affect its welfare?

7. Although the polar bear is not yet extinct, its natural habitat is warming and the number of polar bears in the wild is dwindling as a result. What should be done to save the polar bear, and to Beth Shapiro's question, is there a value in saving a species from extinction if it can only survive in captivity? Take a look at some species we have already saved but are unable to return to their natural habitats. Explain your answer using evidence from conservation resources.

8. Shapiro says that de-extinction can help humans to right our wrongs – she is no doubt referring to the many ways humans have contributed to the demise of several species. Consider the Northern white rhino, a species poached to near extinction: only two living animals – both female – remain in the world as of 2018. Research current efforts at the Leibniz Institute for Zoo and Wildlife Research in Berlin, Germany, and the San Diego Zoo in California. What are the scientists who are working on these projects currently doing to "undo" the damage already created? Do humans have an obligation to try to bring back a species for which we are almost solely responsible for exterminating?

Animals Bow to Their Mechanical Overlords; from *Frankenstein's Cat*

Emily Anthes

Emily Anthes is a science journalist who has written for many publications, including *Scientific American*, the *New York Times*, *The Atlantic*, and *Nature*. She also holds a master's degree in science writing from MIT. In this excerpt from her book, *Frankenstein's Cat: Cuddling Up to Biotech's Brave New Beasts* (2014), Anthes introduces readers to the multiple and surprising ways scientists are bioengineering insects. Initially published as an article in *Nautilus* magazine, this piece explores how technology now enables humans to infiltrate insect communities in order to monitor and/or control their behavior. This technology, in the right hands, has the potential to benefit species who might need a little human help. As you read, consider who decides when, or if, to intervene and why.

Several years ago, a group of American cockroaches discovered four strangers in their midst. A brief investigation revealed that the interlopers *smelled* like cockroaches, and so they were welcomed into the cockroach community. The newcomers weren't content to just sit on the sidelines, however. Instead, they began to actively shape the group's behavior. Nocturnal creatures, cockroaches normally avoid light. But when the intruders headed for a brighter shelter, the rest of the roaches followed.

What the cockroaches didn't seem to realize was that their new, light-loving leaders weren't fellow insects at all. They were tiny mobile robots, doused in cockroach pheromones and programmed to trick the living critters into following their lead. The demonstration, dubbed the LEURRE project and conducted by a team of European researchers, validated a radical idea—that robots and animals could be merged into a "biohybrid" society, with biological and technological organisms forming a cohesive unit.

A handful of scientists have now built robots that can socially integrate into animal communities. Their goal is to create machines that not only infiltrate animal groups but also influence them, changing how fish swim, birds fly, and bees care for their young. If the research reaches the real world, we may one day use robots to manage livestock, control pests, and protect and preserve wildlife. So, dear furry and feathered friends, creepy and crawly creatures of the world: Prepare for a robo-takeover.

Nature has long been a source of inspiration for roboticists. Engineers commonly borrow from biology to design robots that move in novel ways, yielding machines that undulate like snakes, climb like geckoes, or glide like manta rays. But when LEURRE launched in 2002, the research team aimed to reach beyond bio-mimicry. By then, collective behavior had become a hot topic in both biology and robotics. "We started thinking, we have on one side the animals capable of doing those collective behaviors, and we have on the other side those robots being built by roboticists that are able to replicate that same kind of collective behavior," says José Halloy, the physicist who served as the scientific coordinator of LEURRE while at the Free University of Brussels, Belgium. "If we establish a link between the two, well then obviously we will have a mixed group that will act collectively." What's more, Halloy says, the researchers believed they might be able to use the robots to influence the group's natural behavior.

5 The team started with cockroaches, which aren't as social as, say, ants or bees, but demonstrate some collective behavior. In particular, they shelter together in large groups; a cockroach is much more likely to hang his arthropod hat at a hideaway that's home to other roaches than to strike out on his own for a vacant resting place. Halloy and his colleagues set out to create mobile, insect-like robots, or InsBots, that could manipulate a group of real roaches' decisions about where to rest.

The first step was to create robots that could intermingle with live cockroaches. Fortunately, the machines didn't need to be exact replicas of the insects. "That's good news for roboticists because it's very difficult to copy an animal," says Halloy, who is now at Paris Diderot University in France. "What matters is that the robots are sending the relevant cues to the animal." The trick is to identify the few key social signals—which vary from species to species and can be visual, auditory, olfactory, or tactile—that will allow robots to communicate with their flesh-and-blood counterparts.

Cockroaches identify their comrades by smell, so Halloy and his colleagues wrapped their matchbox-sized InsBots in filter paper soaked in cockroach pheromones. They placed the robots, along with a dozen real roaches, into a circular arena that contained two different shelters, one darker than the other. In earlier tests, the real roaches had revealed a clear preference for darkness, choosing to aggregate in lighter refuges only 27 percent of the time.

> "Their goal is to create machines that not only infiltrate animal groups but also influence them, changing how fish swim, birds fly, and bees care for their young."

When the researchers introduced the InsBots, which were programmed to prefer the lighter shelters, the entire group's preference for light more than doubled—the mixed society of roaches and robots ended up in the lighter shelter 61 percent of the time.

Halloy's team provided a compelling proof of principle, but the researchers dreamed of doing more than bossing cockroaches around. Since their roach results were published in *Science* in 2007, they've been figuring out how to integrate robots into other critter communities.[1] In one follow-up project, Halloy and Francesco Mondada, a roboticist at the Swiss Federal Institute of Technology in Lausanne in Switzerland, chose to work with chickens and exploit the well-documented phenomenon of filial imprinting, in which newly hatched chicks form nearly instantaneous social attachments to their mothers. Since the 1930s, biologists have known that chicks imprint on nearly any moving object they encounter in the early hours of their lives, from a person to a model train. This phenomenon was well documented by the zoologist Konrad Lorenz, who was famously photographed with a trail of ducklings waddling along behind him.

10 Halloy and Mondada knew that their surrogate mother didn't need to look like a bird, but research had shown that it would help if it were approximately the right size, so they built a cylindrical, hen-sized mobile robot. Because studies had also revealed that chicks are more likely to imprint on objects that make noise, and stand out from the background, the researchers equipped the robot with a speaker that beeped and installed several rings of lights that displayed changing color patterns. Nine-hour-old chicks then watched through a Plexiglass wall as the so-called PoulBot traveled back and forth, beeping. After each chick had undergone three of these hour-long imprinting sessions, it was placed into the experimental arena with the PoulBot.

About 60 percent of the 213 chicks tested followed the PoulBot closely as it cruised around the arena. In fact, some of the chicks were so attracted to the robot that the researchers had to wrap the robot in a Plexiglass bumper so the birds wouldn't get caught in the robot's wheels. "We said, 'Wow, they can be squashed by the robot. We are going to kill some of them if this continues,'" Halloy recalls.

With PoulBot leading the way, biologists might learn new things about bird behavior. "By having control of robots that are inside the group," Mondada says, "you can do some experiments on the working principles of those societies." For instance, how does a group of chicks act when only a fraction of the flock strongly imprints on a mother figure? And does following behavior change with age? The researchers are still analyzing their data, but they hope that the PoulBot can help answer these and other questions—and do so in a controlled, easily replicable way. Robots, after all, are the kings of consistency, able to perform the same way in trial after trial.

Several researchers have already begun using biomimetic robots to probe animal behavior. With the help of robots, researchers at The

University of Texas at Austin, for example, demonstrated that female Túngara frogs are attracted not only to the sound of male frogs' calls, as was already known, but also to the sight of their expanding vocal sacs.

Robots might also help us control animal behavior outside the laboratory. Mondada imagines a world where robots shepherd free-range chickens out of their coops in the morning and back at night, for example. Robo-sheep or -cattle could one day lead real herds. (In 1999, a robotic sheepdog named Rover corralled a flock of ducks and drove them toward a designated location. The robot was not designed to pass as a duck, however, or forge a social connection with them, and probably worked simply by scaring the birds.) By tapping into animals' social behavior and channels of communication, farmers may be able to manage their livestock in a more "natural," humane way, Mondada says, without needing to rely on restrictive physical measures, such as electric fences or barbed wire.

For Maurizio Porfiri, an engineer at the New York University Polytechnic School of Engineering who has built a series of robotic fish, protecting wild animals has always been the goal. "My feeling was that all this technology that was inspired by nature, was never utilized in nature itself," Porfiri says. 15

Porfiri hopes to develop robotic fish that guide wild fish away from hazards such as oil spills and dams. His work so far suggests that it's a plausible idea. Using a water tunnel to recreate the conditions fish might face while swimming in a flowing river, he discovered that when a plastic fish robot beats its motorized tail at just the right frequency, live golden shiners swim

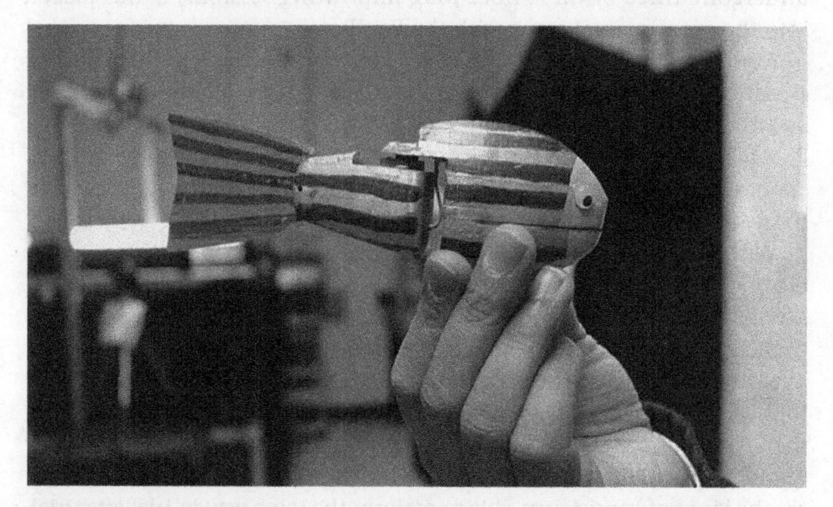

This mechanical zebrafish is designed to attract other zebrafish as well as repel mosquitofish, an invasive species in many waterways.
Nautilus

along behind it. At these frequencies—which vary according to the speed at which the water is flowing—fish in the wake of the robot "obtain a hydrodynamic advantage," Porfiri says, meaning that they need to expend less energy to swim. The finding taught his team how fish robots will need to move in order to be effective leaders in the wild.

Robots may also ward off unwanted species. Over the last several years, Porfiri has developed a mobile robot that resembles a zebrafish; the machine is covered in bright blue and silver stripes, with a few splotches of yellow pigment, and has the slightly rounded shape of a fertile female fish. In a study published last year, Porfiri showed that while the robot attracts zebrafish, it repels mosquitofish, an aggressive invasive species.[2] Over the past century, people deliberately introduced mosquitofish into a variety of new habitats because they eat mosquito larvae. But now, the fish cause significant damage to local amphibian and fish species by eating their eggs and young. In their native habitats, mosquitofish are preyed upon by brightly colored sunfishes, so when the mosquitofish viewed the robot, with its electric blue and yellow paint, it might have perceived it as a predator and been scared away. Similar robots could one day be deployed in the wild to keep invasive species out of critical habitats, Porfiri says.

Now Porfiri is scaling up, trying to understand how to integrate more than one robot into a group of fish. "It's not practical to have one robot controlling a big group," he says. "So you need multiple robots. How should these robots coordinate with each other to influence the behavior of a group? That's a fundamental question."

In Halloy and Mondada's latest endeavor, they are attempting to pull off the same feat with zebrafish as they once did with cockroaches, using a group of biomimetic robots to manipulate the animals' choice of shelter. The work is part of ASSISIbf, a five-year project launched last year that involves scientists in six European countries. (The project is named after St. Francis of Assisi, who reportedly had a gift for communicating with animals.)

In addition to working with zebrafish, the scientists are also developing a network of 64 robots to interact with honeybees. The robots, which are still in development, will generate heat, light, vibrations, and electromagnetic fields—all stimuli known to affect bee behavior. The researchers hope to be able to use a combination of these stimuli to encourage all the bees to aggregate at a single specific location. "As soon as we can do that, we will do other tasks like split up a group of bees into two swarms or four sub-swarms or maneuvering the swarm around," says Thomas Schmickl, a zoologist at the University of Graz, in Austria, who is leading the project. Eventually, Schmickl says, beekeepers may be able to install this kind of robotic system in a bee hive, using it to encourage bees to take more pollination flights, for instance, or to go out looking for food at a certain time of day.

20

Another possibility is to use the robots to manipulate "nurse" bees into taking better care of the colony's brood. If a higher percentage of larvae survive to adulthood, it could help boost the declining bee population and provide more pollinators to tend to our food crops. "There is always a loss in the brood in a honeybee colony, but if you can control the nurses, you can affect the losses, probably," Schmickl says.

In an unrelated project, scientists at the Free University of Berlin in Germany have developed a honeybee robot that can do a basic waggle dance—a complex series of movements bees naturally perform, seemingly to communicate with each other about where food sources are located. Once refined, waggle-dancing robots could potentially be used to drive real bees to fly to specific locations.

It wouldn't be the first time an outsider manipulated bees by pretending to be one of the gang. Certain orchids, for example, trick male bees into pollinating them by displaying petals shaped like beckoning female bees. In comparison, the roboticists' plans are gracious. The grand union of beasts and 'bots may ultimately make animal societies stronger; a little robo-leadership could cause some species' stock to soar.

References

1. Halloy, J., Sempo, G., Caprari, G. Rivault, C., Asadpour, M., et al. 2007. "Social Integration of Robots into Groups of Cockroaches to Control Self-Organized Choices." *Science* 318: 1155–1158.
2. Polverino, G. and Porfiri, M. 2013. Zebrafish (*Danio rerio*) behavioral response to bioinspired robotic fish and mosquitofish (*Gambusia affinis*). *Bioinspired Biomimetics* 8: 044001.

Understanding the Text

1. Although Anthes does not specifically define them, she talks extensively about biomimetic robots. Look up the official definition. What are they? Does this meaning come across in her excerpt even though she does not supply a definition? Why or why not?
2. Throughout this essay, Anthes refers to historical examples and connects them to present-day experiments. What is the purpose of this strategy and how does it add to her credibility as an author, if at all?

Reflection and Response

3. Anthes provides several examples of the potential good that could come from robots integrating with insect and animal communities. For example, she writes that "Robo-sheep or -cattle could one day lead real herds" (par. 14). Think of two other examples. Do you agree that there may be value in these experiments?

4. Based on the information provided and the tone used, identify Anthes's purpose and audience. Use evidence from the excerpt to support your answer.

Making Connections

5. Anthes titled her book *Frankenstein's Cat*, a clear reference to Mary Shelley's novel, but Anthes is just one of many authors who have used a *Frankenstein* reference to write about science. Research other books and articles about science (many are listed in "The Horror Story That Haunts Science," which begins on page 126) that make a connection to the novel. Based on your research, what links do you see between *Frankenstein* and the real world? Why might real scientists consistently reference a fictional book about pseudo-science?
6. This piece was initially published as an article in 2014. Research some of the developments in robotic insects or animals, also known as biomimetic robots, since this essay's publication. What are some of the current uses of these robots?
7. While some of the biomimetic robots that Anthes describes are being used simply to gather information, others have the potential to change the behavior of the animal or insect group they are part of, potentially saving the species. For instance, she writes, "Eventually, Schmickl says, beekeepers may be able to install this kind of robotic system in a bee hive, using it to encourage bees to take more pollination flights" (par. 20). Research a single struggling animal or insect group that you feel may benefit from human intervention. Write an essay justifying or opposing human intervention with the use of robots and exploring the potential risks and benefits in doing so.

The Future of Farming Is Looking Up

Selina Wang

Selina Wang is a reporter for Bloomberg News, where she has written numerous articles on technology and the economy. Wang also frequently contributes on-air to Bloomberg Television's segments on technology. She has a degree in economics from Harvard University.

In this article, from *Bloomberg Businessweek*, Wang examines a new form of farming that has potential to change the way we grow and buy our food. She looks at Plenty Inc., an startup indoor farming business, and its two entrepreneurial owners. While their model, the indoor vertical farm, is fairly new, it is not radically different than some other indoor farming ideas, but the problems it addresses might surprise you. Wang helps us to better understand why this type of farming might be more impactful in certain communities than in others, and how growing indoors addresses some environmental concerns.

Before stepping into Plenty Inc.'s indoor farm on the banks of San Francisco Bay, make sure you're wearing pants and closed-toe shoes. Heels aren't allowed. If you have long hair, you should probably tie it back.

Your first stop is the cleaning room. Open the door and air will whoosh behind you, removing stray dust and contaminants as the door slams shut. Slide into a white bodysuit, pull on disposable shoe covers, and don a pair of glasses with colored lenses. Wash your hands in the sink before slipping on food-safety gloves. Step into a shallow pool of clear, sterilized liquid, then open the door to what the company calls its indoor growing room, where another air bath eliminates any stray particles that collected in the cleaning room.

The growing room looks like a strange forest, with pink and purple LEDs illuminating 20-foot-tall towers of leafy vegetables that stretch as far as you can see. It smells like a forest, too, but there's no damp earth or moss. The plants are growing sideways out of the columns, which bloom with Celtic crunch lettuce, red oak kale, sweet summer basil, and 15 other heirloom munchables. The 50,000-square-foot room, a little more than an acre, can produce some 2 million pounds of lettuce a year.

Step closer to the veggie columns, and you'll spot one of the roughly 7,500 infrared cameras or 35,000 sensors hidden among the leaves. The sensors monitor the room's temperature, humidity, and level of carbon dioxide, while the cameras record the plants' growing phases. The data stream to Plenty's botanists and artificial intelligence experts, who regularly tweak the environment to increase the farm's productivity and enhance the food's taste. Step even closer to the produce, and you may see a ladybug or two. They're there to eat any pests that somehow make it past the cleaning room. "They work for free so we don't have to eat pesticides," says Matt Barnard, Plenty's chief executive officer.

5 Barnard, 44, grew up on a 160-acre apple and cherry orchard in bucolic Door County, Wis., a place that attracts a steady stream of fruit-picking tourists. Now he and his four-year-old startup aim to radically change how we grow and eat produce. The world's supply of fruits and vegetables falls 22 percent short of global nutritional needs, according to public-health researchers at Emory University, and that shortfall is expected to worsen. While the field is littered with the remains of companies that tried to narrow the gap over the past few years, Plenty seems the most promising of any so far, for two reasons. First is its technology, which vastly increases its farming efficiency—and, early tasters say, the quality of its food—relative to traditional farms and its venture-backed rivals. Second, but not least, is the $200 million it collected in July from Japanese telecom giant SoftBank Group, the largest agriculture technology investment in history.

"The growing room looks like a strange forest, with pink and purple LEDs illuminating 20-foot-tall towers of leafy vegetables that stretch as far as you can see."

With the backing of SoftBank CEO Masayoshi Son, Plenty has the capital and connections to accelerate its endgame: building massive indoor farms on the outskirts of every major city on Earth, some 500 in all. In that world, food could go from farm to table in hours rather than days or weeks. Barnard says he's been meeting with officials from some 15 governments on four continents, as well as executives from Wal-Mart Stores Inc. and Amazon.com Inc., while he plans his expansion. (Bezos Expeditions, the Amazon CEO's personal venture fund, has also invested.) He intends to open farms abroad next year; this first one, in the Bay Area, is on track to begin making deliveries to San Francisco grocers by the end of 2017. "We're giving people food that tastes better and is better for them," Barnard says. He says that a lot.

Plenty acknowledges that its model is only part of the solution to the global nutrition gap, that other novel methods and conventional farming will still be needed. Barnard is careful not to frame his crusade in opposition to anyone, including the industrial farms and complex supply chain he's trying to circumvent. He's focused on proving that growing rooms such as the one in South San Francisco can reliably deliver Whole Foods quality at Walmart prices. Even with $200 million in hand, it won't be easy. "You're talking about seriously scaling," says Sonny Ramaswamy, director of the National Institute of Food and Agriculture, the investment arm of the U.S. Department of Agriculture. "The question then becomes, are things going to fall apart? Are you going to be able to maintain quality control?"

The idea of growing food indoors in unlikely places such as warehouses and rooftops has been hyped for decades. It presents a compelling solution to a series of intractable problems, including water shortages, the scarcity of arable land, and a farming population that's graying as young people eschew the agriculture industry in greater numbers. It also promises to reduce the absurd waste built into international grocery routes. The U.S. imports some 35 percent of fruits and vegetables, according to Bain & Co., and even leafy greens, most of which are produced in California or Arizona, travel an average of 2,000 miles before reaching a retailer. In other words, vegetables that are going to be appealing and edible for two weeks or less spend an awful lot of that time in transit.

So far, though, vertical farms haven't been able to break through. Over the past few years, early leaders in the field, including PodPonics in Atlanta, FarmedHere in Chicago, and Local Garden in Vancouver, have shut down. Some had design issues, while others started too early, when hardware costs were much higher. Gotham Greens in Brooklyn, N.Y., and AeroFarms in Newark, N.J., look promising, but they haven't raised comparable cash hoards or outlined similarly ambitious plans.

While more than one of these companies was felled by a lack of expertise in either farming or finance, Barnard's unusual path to his Bay Area warehouse makes him especially suited for the project. He chose a different life than the orchard, frustrated with the degree to which his life could be upended by an unexpected freeze or a broken-down tractor-trailer. Eventually he became a telecommunications executive, then a partner at a private equity firm. In 2007, two decades into his white-collar life, he started his own company, one that concentrated on investing in technologies to treat and conserve water. After an investor suggested he consider putting money into vertical farming, Barnard began to research the subject and quickly found himself obsessed with shortages of food and arable land. "The length of the supply chain, the time and distance it takes," he says, meant "we were throwing away half of the calories we grow." He spent months chatting with farmers, distributors, grocers, and, eventually, Nate Storey. 10

The grandson of Montana ranchers, 36-year-old Storey spent much of his childhood planting and tending gardens with his six siblings. Their Air Force dad, who eventually retired as a lieutenant colonel, moved them to another base every few years, and the family gardened to save money on groceries. "I was always interested in ranching and family legacy but frustrated on how to do it," Storey says. "If you're an 18-year-old kid and you want to farm or ranch, most can't raise $3 million to buy a farm or a ranch."

A decade ago, as a student at the University of Wyoming, he learned about the same industry-level inefficiencies Barnard observed. He began experimenting with vertical farming for his doctoral dissertation in agronomy and crop science, and in 2009 patented a growing tower that would pack the plants more densely than other designs. He spent $13,000, then a sizable chunk of his life savings, to buy materials for the towers and started building them in a nearby garage. By the time he met Barnard in 2013, he'd sold a few thousand to hobbyist farmers and the odd commercial grower.

Storey became Barnard's co-founder and Plenty's chief science officer, splitting his time between Wyoming and San Francisco. Together they made Storey's designs bigger, more efficient, and more readily automated. By 2014 they were ready to start building the farm.

Most vertical farms grow plants on horizontal shelves stacked like a tall dresser. Plenty uses tall poles from which the plants jut out horizontally. The poles are lined up about 4 inches from one another, allowing crops to grow so densely they look like a solid wall. Plenty's setups don't use any soil. Instead, nutrients and water are fed into the top of the poles, and gravity does much of the rest of the work. Without horizontal shelves, excess heat from the grow lights rises naturally to vents in the ceiling. "Because we work with physics, not against it, we save a lot of money," Barnard says.

15 Water, too. Excess drips to the bottom of the plant towers and collects in a recyclable indoor stream, and a dehumidifier system captures the condensation produced from the cooling hardware, along with

A vertical farm wall provides plants with plenty of room to spread out in all directions while simultaneously maximizing "farmed" space.

Brandon Wade/AP Images for Eden Green

moisture released into the air by plants as they grow. All that accumulated H_2O is filtered and fed back into the farm. All told, Plenty says, its technology can yield as much as 350 times more produce in a given area as conventional farms, with 1 percent of the water. (The next-highest claim, from AeroFarms, is as much as 130 times the land efficiency of traditional models.)

Based on readings from the tens of thousands of wireless cameras and sensors, and depending on which crop it's dealing with, Plenty's system adjusts the LED lights, air composition, humidity, and nutrition. Along with that hardware, the company is using software to predict when plants should get certain resources. If a plant is wilting or dehydrated, for example, the software should be able to alter its lighting or water regimen to help.

Barnard, tall and lanky with a smile that crinkles his entire face, becomes giddy when he recounts the first time Plenty built an entire growing room. "It had gone from pretty sparse to a forest in about a week," he says. "I had never seen anything like that before."

When he and Storey started collaborating, their plan was to sell their equipment to small growers across the country. But to make a dent in the produce gap, they realized they'd need to reproduce their model farm with consistency and speed. "If it takes you two or three years to build a facility, forget about it," Storey says. "That's just not a pace that's going to have any impact." That meant they'd have to engineer the farms themselves. And that meant two things: They'd need more than their 40 staffers, and they'd need way more money.

It wasn't easy for Barnard to get his first meeting with Son, in March. One of Plenty's early investors had to beg the SoftBank CEO, who allotted Barnard 15 minutes. He and the investor, David Chao of DCM Ventures, jammed one of the 20-foot grow towers into Chao's Mercedes sedan and took off for Son's mansion in Woodside, Calif., some 30 miles from San Francisco. Son looked bewildered as they unloaded the tower, but the meeting stretched to 45 minutes, and two weeks later they flew to Tokyo for a more official discussion in SoftBank's boardroom. The $200 million investment, announced in late July, will help Plenty put a farm in every major metro area with more than 1 million residents, according to Barnard. He says each will have a grow room of about 100,000 square feet, twice the size of the Bay Area model, and can be constructed in under 30 days.

20 Chao says SoftBank wants "to help Plenty expand very quickly, particularly in China, Japan, and the Middle East," which all struggle with a lack of arable land. Other places on the near-term list include Canada, Denmark, and Ireland. Plenty is also in talks with insurers and institutional investors such as pension funds to bankroll its farm-building with debt. Barnard says

the farms would be able to pay off investors in three to five years, vs. 20 to 40 years for traditional farms. Think of it more like a utility, he says.

Plenty, of course, isn't as sure a bet as Consolidated Edison Inc. or Italy's Enel SpA. The higher costs of urban real estate, and the electricity needed to run all of the company's equipment, cut into its efficiency gains. While it's adapting its technology for foods including strawberries and cucumbers, the complications of tree-borne fruits and root crops likewise neutralize the value of its technology. And Plenty has to contend with commercial farms that have spent decades building their relationships with grocers and suppliers and a system that already offers many people extremely low prices for a much wider variety of goods. "What I haven't seen so far in vertical farm technologies is these entities getting very far beyond greens," says Michael Hamm, a professor of sustainable agriculture at Michigan State University. "People only eat so many greens."

Barnard says he's saving way more on truck fuel and other logistical costs, which account for more than one-third of the retail price of produce, than he's spending on warehousing or power. He's also promising that the company's farms will require long-term labor from skilled, full-time workers with benefits. About 30 people can run the South San Francisco warehouse; future models, which will be about two to five times its size, may require several hundred apiece, he says. While robots can handle some of the harvesting, planting, and logistics, experts will oversee the crop development and grocer relationships on-site.

Retailers shouldn't need much convincing, says Mikey Vu, a partner at Bain who studies the grocery business. "Grocers would love to get another four to five days of shelf life for leafy greens," he says. "I think it's an attractive proposition."

Gourmets like Plenty's results, too. Anthony Secviar, a former sous-chef at the French Laundry, a Michelin-starred restaurant in the Napa town of Yountville, says he wasn't expecting much when he received a box of Plenty's produce at his home in Mountain View, Calif. The deep green of the basil and chives hit him first. Each was equally lush, crisp, flavorful, and blemish-free. "I've never had anything of this quality," says Secviar, who while at the French Laundry cooked with vegetables grown across the street from the restaurant. He's now on Plenty's culinary council and is basing his next restaurant's menu around the startup's heirloom vegetables. "It checks every box from a chef's perspective: quality, appearance, texture, flavor, sustainability, price," he says.

At the South San Francisco farm, the greens are fragrant and sweet, the kale is free of store-bought bitterness, and the purple rose lettuce carries a strong kick. There's enough spice and crunch that the veggies won't need a ton of dressing. Although Plenty bears little resemblance to a quaint family farm, the tastes bring me back to the tiny vegetable patch my grandparents planted in my childhood backyard. It's tough to believe these spicy mustard greens and fragrant chives have been re-created in a sterile room, without soil or sun.

Understanding the Text

1. The "veggie columns" include "roughly 7,500 infrared cameras or 35,000 sensors hidden among the leaves" (par. 4). Where does this data go and who is tracking it?
2. What background do Plenty's founders, Matt Barnard and Nate Storey, share? Why is this important?
3. There are some disadvantages to farming indoors. What are two potential downsides?

Reflection and Response

4. Barnard seems to understand that at its current scale, vertical farming does not yet compete with industrial farms and the "complex supply chain he's trying to circumvent" (par. 7). Why does he believe this? Do you think Barnard is correct about this assumption?
5. One of the selling points of vertical farming is that the food can be grown closer to grocery stores and restaurants, resulting in nearly another week of shelf-life for some vegetables (Wang). Consider your own experiences shopping for produce. Have you purchased produce that was already too ripe to use or not yet ready to eat? How might vertical farming impact you and would you purchase food grown this way?

Making Connections

6. What connections do you see between this article and "Quenching Society's Thirst" (p. 50)? How does each method, desalination and vertical farming, address similar problems?
7. As Wang writes, "Plenty acknowledges that its model is only part of the solution to the global nutrition gap" (par. 7). What does "global nutrition gap" mean in the context of this article? Do some research to define this problem and who it affects. Then, consider whether you think Plenty's model is a feasible solution to partially addressing this problem? What are some other potential solutions?
8. Think about the environment of the vertical farm workplace: what types of skills will employees need to work in a Plenty facility? Do some research in order to compare and contrast these skills with those that traditional farmers need. Then, go a step further – will humans always be necessary for this type of job? Could robots replace these vertical farmers? Why or why not?

Treading the Fine Line between Climate Talk and Alarmism

Sarah E. Myhre

Sarah Myhre is a paleoceanographer, as well as a scientific and social activist. Her research into the history of oceans and their connections to climate change, as well as her vocal activism in support of female scientists, has garnered her several academic awards and appointments, including a Kavli Fellowship. She is currently working on the Future of Ice Initiative at the University of Washington.

In this op-ed, which appeared on LiveScience.com in 2017, Myhre discusses the challenges of climate change while focusing primarily on how nonscientists understand these challenges. Myhre's approach in this op-ed includes careful consideration of how her own social and professional positions affect the rhetorical choices she makes in her effort to communicate effectively with her audience. She acknowledges the complexity of climate change research and considers what else scientists can do to help the general public make sense of this very complicated, often misunderstood, and highly politicized topic.

In May 2017, I spoke about climate change, something I had done often, but this was unique. It was the first time I spoke about the issue with a faith-based community. The talk was a contribution to a springtime "Earth Care" ministry series. I dressed conservatively, and I brought only an activity I use for educating kids at a science museum in Seattle—a hands-on lesson in stratigraphy, superposition and geologic time. No slide deck. No computer. No data. I came to talk about climate change, Earth's history, and public trust and decision-making around the issue.

My prepared remarks quickly were tossed aside, as my presentation became a conversation with the 20 church members. We talked about values, our love of the Pacific Northwest, our shared commitment to steward the Earth and care for those in most need. I spoke about my views as a scientist—about the risks of unchecked greenhouse gas pollution to our planet, our home, and to future generations.

I confessed to the group that my politics often lean left of center, and yet my grandparents, particularly my maternal grandfather, were conservative. So, I value the role of conservative voices in American politics, and I identify with those voices. And yet, the acceptance of the basic science of climate change has cleaved across partisan lines—a political reality that would have my grandfather, a construction engineer and businessman, aghast and angered.

"We are living through a crisis of trust between the American public and climate scientists, and we must extend ourselves, as scientists and public servants, to rebuild transparency and trust with the public."

I am a fifth-generation Washingtonian. While I do not lay claim to the identity of the Pacific Northwest, which, frankly, should be reserved for the peoples of, say, the Tulalip Tribe and the Nooksack Tribe, I do identify with and love this land of mountaintop archipelagos, cold rivers and steep, deep skiing. We in the Pacific Northwest are not exempt from the physical disruptions that come with climate change.

As an example, the city of Seattle is planning for average annual temperatures to increase within a range of 1.5 to 5.2 degrees Fahrenheit (0.8 to 3 degrees Celsius) by the 2040s, with summer temperatures increasing by as much as 7.9 degrees Fahrenheit (4.4 degrees Celsius), according to the Seattle Climate Action Plan. And neighboring Vancouver can expect summer temperatures by the 2050s to be somewhere between those of present-day Seattle and San Diego. 5

Now, we, *collectively,* need to make decisions around the highest temperature projections. This is because, when we talk about carbon emission scenarios and climate sensitivity, we are ultimately talking about future risk management. The highest cost in public health and public resources will come with risk associated with the warmest possible futures—and this should be where we focus our attention.

Climate concerns are not just about temperature. Big pieces of the Earth's system also change when we alter the global carbon cycle through adding a heat-trapping greenhouse gas "blanket" to the atmosphere. For us in the Pacific Northwest, this means that our snowpack and mountain recreation lifestyles are vulnerable; our rivers of salmon and eagles are vulnerable; and our cold coastlines and marine economies are vulnerable. Put simply, our water and our people are at risk.

We have a lot to lose in the face of unchecked climate warming. Not to be too personal, but have you been to San Diego lately? I would be a different person had I grown up in the heat and glamor of Southern California, rather than in cold, dark, rainy Seattle.

As my talk came to its end, a quiet man in his mid-50s spoke up, slowly and measuredly. He told me, "You know, no one wants to be called an alarmist. But it is OK to sound the alarm on this."

I heard this man's kind words and slumped back into my chair, my heart pierced by this plain-spoken and unvarnished advice. I have chosen to walk a tightrope as a public scholar, by turning toward the immense challenge of communicating the terrifying and heartbreaking (and I mean those words specifically) risks that come with climate change. 10

Like most scientists, the last thing that I want is to be called is an alarmist. To be an alarmist smacks of everything we are trained to avoid as academics—ideology, magical thinking, self-inflation, ego (to be sure, I am still working on all these pieces). This advice from a stranger in a church in Everett, Washington, vented a pressure valve in my mind—this impossible bind between communicating alarming information and deeply eschewing the "alarmist" public role. The late Steve Schneider wrote about the double ethical bind of communicating both effectively and honestly as a scientist, and described it as "no-win scenario." Based on my experience in the public eye, and specifically as a female academic, I agree.

What is our role in public leadership as scientists? I would suggest a few action items: Work to reduce risk and cost for the public; steward the public's interest in evidence; and be steady and committed to the scientific process of dissent, revision and discovery. This means communicating risk when necessary. We would never fault an oncologist for informing patients about the cancer risks that come with smoking. Why would we expect Earth scientists to be any different, when we're just as certain?

As a public scholar with expertise in paleoclimate science, I communicate alarming, difficult information about the consequences to Earth and ocean systems that have come with past events of abrupt climate warming. As the saying goes, *the past is the key to the future.*

Here is the rub about being a trusted public source of information—you cannot just be a content expert. You must also be a person. To earn trust in the public eye, you have to disclose your conflicts of interest. You must embrace transparency. You must articulate the limits of your expertise. You must come to see the line separating evidence and your own ideology. And I think this transparency made it possible for me to build trust with a suburban community of faith—to talk about this truly alarming information.

The challenge is—how do we do this work better? As scientists, we must build a coherent, evidence-based communication plan to participate in public dialogue across an acrimonious, partisan, human landscape—because it is a shark tank out there, especially for younger, untenured (and marginalized) academics.

We are living through a crisis of trust between the American public and climate scientists, and we must extend ourselves, as scientists and public servants, to rebuild transparency and trust with the public. I will start: I want the global community to mitigate the extreme risk of the warmest future climate scenarios. And, I want my kid to eat salmon and ski with his grandkids in the future. I am invested in that cooler, safer, more sustainable future—for your kids and for mine. Just don't call me an alarmist.

Understanding the Text

1. When considering temperature changes, Myhre says we are really considering "future risk management" (par. 6). What does she mean by this? What is risk management, and how do you think the concept applies to global temperatures?
2. Myhre says that she doesn't want to be considered "an alarmist" (par. 11). As a scientist, why is Myhre wary of being seen sounding the alarm on this issue?
3. In her final paragraph, Myhre discusses how a scientist can build trust with the public, in part by being honest about who he or she is as a person. Did Myhre follow her own advice in this article? What did you learn about her as a person in addition to learning about her as a scientist? What effect did that information have on you, a member of her audience the general public?

Reflection and Response

4. At the beginning of her op-ed, Myhre tells us that she was speaking to a faith-based community and that she told the members that her own political leanings were "left of center." What assumptions do you think she may have made in advance about her audience? What approaches did she identify or use to communicate with this audience?
5. This piece is an op-ed. What is an op-ed? How does it differ from a scientific article? Why do you think it is important to understand what characterizes this genre as you analyze Myhre's writing?

Making Connections

6. Climate change is not only an environmental issue; it is a controversial political issue as well. As a result, there is a lot of misinformation published about climate change, so determining what is true and not true can sometimes be challenging. Find three articles that support climate change and three articles that argue against it. Then, highlight any information that you find in more than one source. How is the same information conveyed by different authors? Is it used for different purposes?
7. Using Myhre's op-ed as a reference, write about a controversial environmental issue, perhaps even a local issue that affects your own community. Imagine that your audience is not yet convinced of that the issue is a problem. Complete some research on the topic, and use a Rogerian argument format, where you establish common ground with your audience and identify areas where you can come together to work toward a solution, to support your points.
8. In the article "England's HSBC Issues Stark Warning: Earth Is Running out of Resources to Sustain Life" at Forbes.com, science writer Trevor Nace says it is "rare for a bank to chime in on climate change and environmental management, [but] HSBC analysts believe it's essential to include climate risks in future financial models." Compare this claim to Myhre's argument that considering climate change is really a form of risk management. Based on these readings, how is climate change also a financial issue?

Why Geoengineering?; from *Earthmasters: The Dawn of the Age of Climate Engineering*

Clive Hamilton

Clive Hamilton, an Australian author and public intellectual, has devoted a large swath of his career to writing about climate change and the politics that surround it. Hamilton has written over fifteen books and has held visiting academic positions at various universities around the world. Currently, he is a professor of public ethics at Charles Sturt University in Canberra, Australia.

In this excerpt from his book *Earthmasters: The Dawn of the Age of Climate Engineering* (2013), Hamilton articulates the consequences of an increasingly warmer planet and examines potential solutions that geoengineers° have posited. One such proposed solution is to mine dust on the moon and scatter it in a "particle cloud" between the Earth and the sun in order to block solar rays. These are drastic solutions to a dire problem, and while they may sound extreme, as Hamilton explains, scientists fear that the public doesn't have a grasp on just how dire the threat of climate change has become.

Climate Fix

As the effects of global warming begin to frighten us, geoengineering will come to dominate global politics. Scientists and engineers are now investigating methods to manipulate the Earth's cloud cover, change the oceans' chemical composition and blanket the planet with a layer of sunlight-reflecting particles. Geoengineering—deliberate, large-scale intervention in the climate system designed to counter global warming or offset some of its effects—is commonly divided into two broad classes. Carbon dioxide removal technologies aim to extract excess carbon dioxide from the atmosphere and store it somewhere less dangerous. This approach is a kind of clean-up operation after we have dumped our waste into the sky. Solar radiation management technologies seek to reduce the amount of sunlight reaching the planet, thereby reducing the amount of energy trapped in the atmosphere of "greenhouse Earth." This is not a clean-up but an attempt to mask one of the effects of dumping waste into the sky, a warming globe.

Diligent contributors to Wikipedia have listed some 45 proposed geoengineering schemes or variations on schemes. Eight or ten of them are receiving serious attention. Some are grand in conception, some are prosaic; some

Geoengineers: scientists who develop large-scale processes to address climate change through geographical manipulation.

"Geoengineering—deliberate, large-scale intervention in the climate system designed to counter global warming or offset some of its effects"

are purely speculative, some are all too feasible; yet all of them tell us something interesting about how the Earth system works. Taken together they reveal a community of scientists who think about the planet on which we live in a way that is alien to the popular understanding. Let me give a few examples.

It is well known that as the sea-ice in the Arctic melts the Earth loses some of its albedo or reflectivity—white ice is replaced by dark seawater which absorbs more heat. If a large area of the Earth's surface could be whitened then more of the Sun's warmth would be reflected back into space rather than absorbed. A number of schemes have been proposed, including painting roofs white, which is unlikely to make any significant difference globally. What might be helpful would be to cut down all of the forests in Siberia and Canada. While it is generally believed that more forests are a good thing because trees absorb carbon, boreal (northern) forests have a downside. Compared to the snow-covered forest floor beneath, the trees are dark and absorb more solar radiation. If they were felled the exposed ground would reflect a significantly greater proportion of incoming solar radiation and the Earth would therefore be cooler. If such a suggestion appears outrageous it is in part because matters are never so simple in the Earth system. Warming would cause the snow on the denuded lands to melt, and the situation would be worse than before the forests were cleared.

More promisingly perhaps, at least at a local scale, is the attempt to rescue Peruvian glaciers, whose disappearance is depriving the adjacent grasslands and their livestock of their water supply. Painting the newly dark mountains with a white slurry of water, sand and lime keeps them cooler and allows ice to form; at least that is the hope.[1] The World Bank is funding research.

Another idea is to create a particle cloud between the Earth and the Sun from dust mined on the moon and scattered in the optimal place.[2] This is reminiscent of the U.S. military's "black cloud experiment" of 1973, which simulated the effect on the Earth's climate of reducing incoming solar radiation by a few percent.[3] Consistent with the long history of military interest in climate control, the study was commissioned by the Defense Advanced Research Projects Agency, the Pentagon's technology research arm, and carried out by the RAND Corporation, the secretive think tank described as "a key institutional building block of the Cold War American empire."[4] I summon up the black cloud experiment here to flag the nascent military and strategic

5

interest being stirred by geoengineering. As we will see in chapter 5, the attention of the RAND Corporation has recently returned to climate engineering. [. . .]

In 1993 the esteemed journal *Climatic Change* published a novel scheme to counter global warming by the Indian physicist P. C. Jain.[5] Professor Jain began by reminding us that the amount of solar radiation reaching the Earth varies in inverse square to the distance of the Earth from the Sun. He therefore proposed that the effects of global warming could be countered by increasing the radius of the Earth's orbit around the Sun. An orbital expansion of 1–2 percent would do it, although one of the side effects would be to add 5.5 days to each year. He then calculated how much energy would be needed to bring about such a shift in the Earth's celestial orbit. The answer is around 10^{31} joules. How much is that? At the current annual rate of consumption, it is more than the amount of energy humans would consume over 10^{20} years, or 100 billion billion years (the age of the universe is around 14 billion years). This seems like a lot, yet Professor Jain reminds us that "in many areas of science, seemingly impossible things at one time have become possible later." Perhaps, he speculates, nuclear fusion will enable us to harness enough energy to expand the Earth's orbit. He nevertheless counsels caution: "The whole galactic system is naturally and delicately balanced, and any tinkering with it can bring havoc by bringing alterations in orbits of other planets also."[6]

The caution is well taken, although the intricate network of orbital dependence has stimulated another geoengineering suggestion. The thought is to send nuclear-armed rockets to the asteroid belt beyond the planets of our solar system so as to "nudge" one or more into orbits that would pass closer to the Earth. Properly calibrated, the sling-shot effect from the asteroid's gravity would shift the Earth orbit out a bit.[7] Of course, if the calibration were a little out, the planet could be sent careening off into a cold, dark universe, or suffer a drastic planet-scale freezing from the dust thrown up by an asteroid strike.

Some of these schemes seem properly to belong in an H. G. Wells° novel or a geeks' discussion group, and too much emphasis on them for the delights of ridicule would give a very unbalanced impression of the research program into climate engineering now underway. That imbalance will be rectified in the next chapters where we will see that serious work is being conducted on schemes to regulate the Earth system by changing the chemical composition of the world's oceans, modifying the layer of clouds that covers a large portion of the oceans and installing

H. G. Wells: (1866–1946), British science fiction writer.

a "solar shield," a layer of sulfate particles in the upper atmosphere to reduce the amount of sunlight reaching the planet. There are some who believe that we will have no choice but to resort to these radical interventions. How did we get to this point? The simple answer is that the scientists who understand climate change most deeply have become afraid.

Hope Against Fear

In 1959 David E. Price, MD, U.S. Assistant Surgeon General, addressed a conference of industrial hygienists with these words:

We live under the shadow of a haunting fear that something may corrupt the environment to the point where man joins the dinosaurs as an obsolete form of life. And what makes these thoughts all the more disturbing is the knowledge that our fate could perhaps be sealed 20 or more years before the development of symptoms.[8]

The shadow under which Americans lived was the dual fear of atomic radiation and chemical pollution. Trepidation that the air might be unsafe to breathe gripped the nation. It was the not-knowing that gave rise to a "mass investment in worry" unmatched, said Price, by an investment in efforts to find out. All that was to change within a few years, spurred by Rachel Carson's° earth-shaking book *Silent Spring,* published in 1962, which both confirmed American anxieties about the impact of the chemical war in agriculture and triggered the rise of modern environmentalism.

The haunting fear that something is corrupting the environment has returned, at least for some. Within our breasts fear and hope are duelling. For a few, the reasons to be afraid have prevailed; for most, hope fights on valiantly. Yet hope wages a losing battle; as the scientists each month publish more reasons to worry, and the lethargy of political leaders drains the wellsprings of hope. In 1959 Dr. Price invoked that all-conquering sentiment of American greatness, unbounded optimism: "Stronger than fear is the conviction that what may at times appear to be the shadow of extinction is in reality the darkness preceding the dawn of the greatest era of progress man has ever known."[9] He was right about the post-war decades. But the world has changed, and now there is a constant trickle of defectors, traitors to hope. To pick out one, the chair of the International Risk Governance Council, Donald Johnston, for ten years the secretary-general of the Organisation for Economic Co-operation and Development

10

Rachel Carson: (1907–1964), American conservationist who argued that humans should not try to control nature with science, primarily chemicals.

(OECD), recently wrote: "By nature I am not a pessimist, but it requires more optimism than I can generate to believe" that the world will limit warming to 2°C higher than the pre-industrial level.[10] Business as usual is a more likely scenario, he added, taking the concentration of carbon dioxide in the atmosphere from its pre-industrial level of 280 parts per million past its current 395 ppm to 700 ppm this century, "with horrendous climate change and unthinkable economic and societal consequences."

The anxiety deepened each year through the 2000s as it became clearer that the range of emissions paths mapped out by experts in the 1990s were unduly optimistic and that the actual growth in emissions, boosted by explosive growth in China, has described a pathway that is worse than the worst-case scenario. When scientists announced that the growth of global greenhouse gas emissions in 2010 was almost 6 percent, breaking all previous records and wiping out the benefits of a temporary lull due to the global recession, many climate scientists around the world drew a sharp in-breath.

The International Energy Agency of the OECD is a staid organization that for years has shared the worldview of oil and coal industry executives. It is the last international body that could be accused of green sympathies, other than the Organization of Petroleum Exporting Countries. So a frisson of dread ran through the climate change community in November 2011 when the IEA released its annual *World Energy Outlook,* the "bible" of the energy sector. It exposed the target of keeping warming below the "dangerous" level of 2°C as a pipe-dream; on current projections, the energy infrastructure expected to be in place as early as 2017 will be enough to lock in future carbon emissions that will warm the Earth by much more. Coal-fired power plants have a lifetime of 50 or 60 years. Waiting for new energy technologies is not an option. If governments do no more than implement the policies they are currently committed to, the IEA expects the world to warm by 3.5°C by the end of the century. "On planned policies, rising fossil energy use will lead to irreversible and potentially catastrophic climate change."[11] If those policy goals prove to be more aspirational than actual then the world is on track for average warming of 6°C above pre-industrial levels, which is almost unthinkable.

It's hard to communicate to the public what a world warmed by 3.5°C will be like, let alone 6°C, or even that the IEA, and all the other organizations saying the same thing, should be taken seriously.[12] After all, for many people one unseasonable snowstorm is enough to nullify decades of painstaking scientific study. And psychologists have discovered that, after accounting for all other factors, when people are put in a room and asked about climate change they are significantly more likely to agree that global warming is "a proven fact" if the thermostat is turned up.[13] Patients with diseases they believe to be serious but untreatable are markedly less likely to agree to diagnostic tests.[14] If it's bad, I don't want to know. Suffice it to say here that 3.5°C means a different kind of world, one hotter than it has been for 15 million years, and not the kind of world on which modern life forms evolved. It would be, eventually, a world without ice—no glaciers, no Arctic sea-ice, no Greenland ice sheet and, almost inconceivably, no Antarctic ice mass. The destabilization of the Earth's climate and natural systems expected this century under the IEA's more "optimistic" scenario would cascade through the centuries beyond.

Notes

1. Rafael Romo, 'Whitening mountains: A new effort to save Peruvian Andes glaciers', *CNN U.S.*, 28 Nov. 2011, at https://edition.cnn.com/2011/11/28/world/americas/peru-mountain-whitening/index.html (accessed Jan. 2012).
2. 'Keep Earth cool with moon dust', *New Scientist,* 9 Feb. 2007, at http://www.newscientist.com/article/dn11151-keep-earth-cool-with-moon-dust.html (accessed Jan. 2012).
3. A. B. Kahle and D. Deirmendjian, 'The black cloud experiment', Report R-1263-ARPA, RAND Corporation, Santa Monica, CA.
4. Chalmers Johnson, 'A litany of horrors: America's university of imperialism', 29 Apr. 2008, at http://www.tomdispatch.com/post/174925/chalmers_Johnson_teaching_imperialism_101 (accessed Jan. 2012). Johnson was once a consultant to RAND.
5. P. C. Jain, 'Earth-Sun system energetics and global warming', *Climatic Change,* 24 (1993), pp. 271–72.
6. Ibid., p. 272.
7. See https://groups.google.com/group/geoengineering/browse_thread/thread/660846de67b3a26c (accessed Jan. 2012). Geoengineering advocate Andrew Lockley describes asteroid nudging as 'a fun idea'.
8. D. Price, 'Is man becoming obsolete?', *Public Health Reports,* 74 (Aug. 1959), p. 693.
9. Ibid., p. 694.
10. Foreword to G. Morgan and K. Ricke, 'Cooling the Earth through solar radiation management: The need for research and an approach to governance', opinion piece for International Risk Governance Council, 2010.
11. International Energy Agency, *World Energy Outlook 2011* (Paris: IEA, 2011), p. 2.
12. The best attempt is by Mark Lynas in *Six Degrees: Our Future on a Hotter Planet* (London: Harper Perennial, 2007).
13. Jane Risen and Clayton Critcher, 'Visceral fit: While in a visceral state, associated states of the world seem more likely', *Journal of Personality and Social Psychology,* 100:5 (2011).
14. Erica Dawson and Kenneth Savitsky, '"Don't tell me, I don't want to know": Understanding people's reluctance to obtain medical diagnostic information', *Journal of Applied Social Psychology,* 36:3 (2006).

Understanding the Text

1. Hamilton explains that "it's hard to communicate to the public what a world warmed by 3.5°C might look like" (par. 13). What makes it so difficult? What evidence does he provide?
2. According to Hamilton, there is a gap between what scientists see as necessary to avoid climate disaster and what governments are actually doing to address the problem of climate change. What are a couple of the gaps he discusses? Explain these in your own words.

Reflection and Response

3. Based on the evidence provided and Hamilton's tone, what do you feel is his purpose for writing this piece? Who is his audience? Provide examples from the text to support your answer.
4. Consider your own beliefs about climate change prior to reading this article. How did Hamilton's writing confirm or refute what you already knew? Choose two pieces of evidence from the excerpt that you found surprising (regardless of whether you agreed or disagreed with the information) and develop two questions about the evidence that you could research further. Why might these questions be worth exploring?

Making Connections

5. There are a comparatively small number of geoengineers in the world and there is often great skepticism of geoengineering both by scientists and the general public. Research the plan to dim the sun developed by the Stratospheric Controlled Perturbation Experiment (ScoPEx) at Harvard University. Explore that plan's potential strengths, risks, and validity.
6. Research the Biosphere 2 project of the late 1980s and early 1990s. What was its mission and how was that project connected to geoengineering? Using your research, develop an argument about the lessons we might learn from the Biosphere 2 project, especially regarding the challenge to manage carbon dioxide.

Quenching Society's Thirst: Desalination May Soon Turn a Corner, from Rare to Routine

Thomas Sumner

Thomas Sumner is a staff writer at the Simons Foundation, a significant benefactor of scientific grants and research. He has written about the sciences for *LiveScience*, *Fox News*, and the *Daily Mail*. Prior to becoming a journalist, Sumner attended University of California, Santa Cruz, where he studied physics.

In his 2016 article from *Science News*, Sumner explores desalination, the process of removing salt from water, as a possible solution to the lack of fresh water that so many major cities now face as a result of population growth, resource overuse, and drought. Given that approximately 97 percent of the Earth's surface is covered in oceans, desalination is an enticing prospect. In 2017 Cape Town, South Africa, a city of 4 million people, declared that it would officially run out of water by July 2018. With drastic rationing and water recycling efforts, the city successfully postponed what the South African government called "Day Zero." However, many other major cities across the globe may also find themselves shutting off the water if alternative sources are not explored. Might desalination be the best way to address our freshwater shortages?

The world is on the verge of a water crisis.

Rainfall shifts caused by climate change plus the escalating water demands of a growing world population threaten society's ability to meet its mounting needs. By 2025, the United Nations predicts, 2.4 billion people will live in regions of intense water scarcity, which may force as many as 700 million people from their homes in search of water by 2030.

Those water woes have people thirstily eyeing the more than one sextillion liters of water in Earth's oceans and some underground aquifers with high salt content. For drinking or irrigation, the salt must come out of all those liters. And while desalination has been implemented in some areas—such as Israel and drought-stricken California—for much of the world, salt-removal is a prohibitively expensive energy drain.

Scientists and engineers, however, aren't giving up on the quest for desalination solutions. The technology underlying modern desalination has been around for decades, "but we have not driven it in such a way as to be ubiquitous," says UCLA chemical engineer Yoram Cohen. "That's what we need to figure out: how to make desalination better, cheaper and more accessible."

Recent innovations could bring costs down and make the technology more accessible. A new wonder material may make desalination plants more efficient. Solar-powered disks could also serve up freshwater with no need for electricity. Once freshwater is on tap, coastal floating farms could supply food to Earth's most parched places, one scientist proposes.

5

Watering Holes

Taking the salt out of water is hardly a new idea. In the fourth century B.C., Aristotle noted that Greek sailors would evaporate impure water, leaving the salt behind, and then condense the vapor to make drinkable water. In the 1800s, the advent of steam-powered travel and the subsequent need for water without corrosive salt for boilers prompted the first desalination patent, in England.

Most modern desalination plants use a technique that differs from these earlier efforts. Instead of evaporating water, pumps force pressurized saltwater from the ocean or salty underground aquifers through special sheets. These membranes contain molecule-sized holes that act like club bouncers, allowing water to pass through while blocking salt and other contaminants.

The membranes are rolled like rugs and stuffed into meter-long tubes with additional layers that direct water flow and provide structural support. A large desalination plant uses tens of thousands of membranes that fill a warehouse. This process is known as reverse osmosis and the result is salt-free water plus a salty brine waste product that is typically pumped underground or diluted with seawater and released back into the ocean. It takes about 2.5 liters of seawater to make 1 liter of freshwater.

In 2015, more than 18,000 desalination plants worldwide had the annual capacity to produce 31.6 trillion liters of freshwater across 150 countries. While still less than 1 percent of worldwide freshwater usage, desalination production is two thirds higher than it was in 2008. Driving the boom is a decades-long drop in energy requirements thanks to innovations such as energy-efficient water pumps, improved membranes and plant configurations that use outbound water to help pressurize incoming water. Seawater desalination in the 1970s consumed as much as 20 kilowatt-hours of energy per cubic meter of produced fresh water; modern plants typically require just over three kilowatt-hours.

10

There's a limit, however, to the energy savings. Theoretically, separating a cubic meter of freshwater from two cubic meters of seawater requires a minimum of about 1.06 kilowatt-hours of energy. Desalination is typically only viable when it's cheaper than the next alternative water source, says Brent Haddad, a water management expert at the

"In 2015, more than 18,000 desalination plants worldwide had the annual capacity to produce 31.6 trillion liters of freshwater across 150 countries."

University of California, Santa Cruz. Alternatives, such as reducing usage or piping freshwater in from afar, can help, but these methods don't create more H_2O. While other hurdles remain for desalination, such as environmentally friendly wastewater disposal, cost is the main obstacle.

The upfront cost of each desalination membrane is minimal. For decades, most membranes have been made from polyamide, a synthetic polymer prized for its low manufacturing cost—around $1 per square foot. "That's very, very cheap," says MIT materials scientist Shreya Dave. "You can't even buy decent flooring at Home Depot for a dollar a square foot."

But polyamide comes with additional costs. It degrades quickly when exposed to chlorine, so when the source water contains chlorine, plant workers have to add two steps: remove chlorine before desalination, then add it back later, since drinking water requires chlorine as a disinfectant. To make matters worse, in the absence of chlorine, the membranes are susceptible to growing biological matter that can clog up the works.

With these problems in mind, researchers are turning to other membrane materials. One alternative, graphene oxide, may knock polyamide out of the water.

Membrane Maze

Since its discovery in 2004, graphene has been touted as a supermaterial, with proposed applications ranging from superconductors to preventing blood clots.

Each graphene sheet is a single-atom-thick layer of carbon atoms arranged in a honeycomb grid. As a hypothetical desalination membrane, graphene would be sturdy and put up little resistance to passing water, reducing energy demands, says MIT materials scientist Jeff Grossman. 15

Pure graphene is astronomically expensive and difficult to make in large sheets. So Grossman, Dave and colleagues turned to a cheaper alternative, graphene oxide. The carbon atoms in graphene oxide are bordered by oxygen and hydrogen atoms.

Those extra atoms make graphene oxide "messy," eliminating many of the material's unique electromagnetic properties. "But for a membrane, we don't care," Grossman says. "We're not trying to run an electric current through it, we're not trying to use its optical properties—we're just trying to make a thin piece of material we can poke holes into."

The researchers start with graphene flakes peeled from hunks of graphite, the form of carbon found in pencil lead. Researchers suspend the graphene oxide flakes, which are easy and cheap to make, in liquid. As a vacuum sucks the liquid out of the container, the flakes form a sheet. The researchers bind the flakes together by adding chains of carbon and oxygen atoms. Those chains latch on to and connect the graphene oxide flakes, forming a maze of interconnected layers. The length of these chains is fine-tuned so that the gaps between flakes are just wide enough for water molecules, but not larger salt molecules, to pass through.

The team can fashion paperlike graphene oxide sheets a couple of centimeters across, though the technique should easily scale up to the roughly 40-square-meter size currently packed into each desalination tube, Dave says. Furthermore, the sheets hold up under pressure. "We are not the only research group using vacuum filtration to assemble membranes from graphene oxide," she says, "but our membranes don't fall apart when exposed to water, which is a pretty important thing for water filtration."

The slimness of the graphene oxide membranes makes it much easier for water molecules to pass through compared with the bulkier polyamide, reducing the energy needed to pump water through them. Grossman, Dave and colleagues estimated the cost savings of such highly permeable membranes in 2014 in a paper in Energy & Environmental Science. Desalination of ground water would require 46 percent less energy; processing of saltier seawater would use 15 percent less, though the energy demands of the new prototypes haven't yet been tested. 20

So far, the new membranes are especially durable, Grossman says. "Unlike polyamide, graphene oxide membranes are resilient to important cleaning chemicals like chlorine, and they hold up in harsh chemical environments and at high temperatures." With lower energy requirements and no need to remove and replace chlorine from source water, the new membranes could be one solution to many desalination challenges.

In large quantities, the graphene oxide membranes may be economically viable, Dave predicts. At scale, she estimates that manufacturing graphene oxide membranes will cost around $4 to $5 per square foot—not drastically more expensive than polyamide, considering its other benefits. Existing plants could swap in graphene oxide membranes when older polyamide membranes need replacing, spreading out the cost of the upgrade over about 10 years, Dave says. The team is currently patenting its membrane-making methodology, though the researchers think it will take a few more years before the technology is commercially viable.

"We are at a point where we need a quantum leap, and that can be achieved by new membrane structures," says Nikolay Voutchkov, executive director of Water Globe Consulting, a company that advises industries and municipalities on desalination projects. The work on graphene oxide "is one way to do it."

Other materials are also vying to be polyamide's successor. Researchers are testing carbon nanotubes, tiny cylindrical carbon structures, as a desalination membrane. Which material wins "will come down to cost," Voutchkov says. Even if graphene oxide or other membranes save money in the long run, high upfront costs would make them less appealing.

Plus, those new membranes won't solve the problems of desalination in less-developed areas. The costs of building a large plant and pumping freshwater over long distances make desalination a hard sell in rural Africa and other water-starved places. For hard-to-reach locales, scientists are thinking small.

25

A Portable Approach

In remote Africa, electricity is hard to come by. Materials scientist Jia Zhu of Nanjing University in China and colleagues are hoping to bring drinkable water to unpowered, parched places by turning to an old-school desalination technique: evaporating and condensing water.

Their system runs on sunshine, something that is both free and abundant in Earth's hotter regions. Using the sun's rays to desalinate water is hardly new, but most existing systems are inefficient. Only about 30 to 45 percent of incoming sunlight typically goes into evaporating water, which means a big footprint is needed to create sizable amounts of freshwater. Zhu and colleagues hope to boost efficiency with a more light-absorbing material.

The material's fabrication starts with a base sheet made of aluminum oxide speckled with 300-nanometer-wide holes. The researchers then coat this sheet with a thin layer of aluminum particles.

When light hits aluminum particles inside one of the holes, the added energy makes electrons in the aluminum start to oscillate and ripple. These electrons can transfer some of that energy to their surroundings, heating and evaporating nearby water without the need for boiling.

The researchers have produced 2.5-centimeter–wide disks of the new material so far, which are light enough to float. The black disks absorb more than 96 percent of incoming sunlight and about 90 percent of the absorbed energy is used in evaporating water, the researchers reported in the June Nature Photonics.

30

The evaporated water condenses and collects in a transparent box containing stainless steel. In laboratory tests, the researchers successfully desalinated water from China's Bohai Sea to levels low enough to meet drinking water standards. The researchers reckon that they can produce around five liters of fresh water per hour for every square meter of material under intense light. In early tests, the disks held up after multiple uses without dropping in performance. Aluminum is cheap and the material's fabrication process can easily scale, Zhu says. While the disks can't produce as much drinkable water as quickly as big desalination plants, the new method may serve a different need, since it's more affordable and more portable, he says. "We are developing a personalized water solution without big infrastructure, without extra energy consumption and with a minimum carbon footprint." The researchers hope that their new desalination technique will find use in developing countries and remote areas where conventional desalination plants aren't feasible.

The disks are worth pursuing, says Haddad at UC Santa Cruz. "I say let's try it out. Let's work with some villages and see how well the tech works and get their feedback. That to me is a good next step to take."

Desalinating water by evaporation has a downside, though, Voutchkov says. Unlike most methods for removing salt, evaporation produces pure distilled water without any important dissolved minerals such as calcium and magnesium. Drinking water without those minerals can cause health issues over time, he warns. "It's OK for a few weeks, but you can't drink it forever." Minerals would need to be added back in to the water, which is hard to do in remote places, he says.

Freshwater isn't just for filling water bottles, though. With a nearly endless supply of salt-free water at hand, desalination could bring agriculture to new places.

Coastal Crops

When Khaled Moustafa looks at a beach, he doesn't just see a place for sunning and surfing. The biologist at the National Conservatory of Arts and Crafts in Paris sees the future of farming. 35

In the April issue of Trends in Biotechnology, Moustafa proposed that desalination could supply irrigation water to colossal floating farms. Self-sufficient floating farms could bring agriculture to arid coastal regions previously inhospitable to crops. The idea, while radical, isn't too farfetched, given recent technological advancements, Moustafa says.

Floating farms would lay anchor along coastlines and suck up seawater, he proposes. A solar panel–powered water desalination system would provide freshwater to rows of cucumbers, tomatoes or strawberries stacked like a big city highrise inside a "blue house" (that is, a floating greenhouse).

Each floating farm would stretch 300 meters long by 100 meters wide, providing about 3 square kilometers of cultivable surface over only three-tenths of a square kilometer of ocean, Moustafa says. The farms could even be mobile, cruising around the ocean to transport crops and escape bad weather.

Such a portable and self-contained farming solution would be most appealing in dry coastal regions that get plenty of sunshine, such as the Arabian Gulf, North Africa and Australia.

"I wouldn't say it's a silly idea," Voutchkov says. "But it's an idea that can't get a practical implementation in the short term. In the long term, I do believe it's a visionary idea." 40

Floating farms may come with a large price tag, Moustafa admits. Still, expanding agriculture should "be more of a priority than building costly football stadiums or indoor ski parks in the desert," he argues.

Whether or not farming will ever take to the seas, new desalination technologies will transform the way society quenches its thirst. More than 300 million people rely on desalination for at least some of their daily water, and that number will only grow as needs rise and new materials and techniques improve the process. "Desalination can sometimes get a rap for being energy intensive," Dave says. "But the immediate benefits of having access to water that would not otherwise be there are so large that desalination is a technology that we will be seeing for a long time into the future."

Understanding the Text

1. Sumner states that humans have been taking salt out of water for centuries. What are some of the ways this has been done in the past? Do you think these methods would still work today? Why or why not?
2. According to Sumner, "for much of the world, salt-removal is a prohibitively expensive energy drain" (par. 3). What examples of this cost does Sumner provide?
3. What is the connection between the desalination process and electricity? Why might a lack of electricity be a barrier for some in choosing desalination?

Reflection and Response

4. Sumner begins his essay with the stark declaration "The world is on the verge of a water crisis." Why do you think Sumner begins the article with such a stark declaration? What questions does it bring to mind as you read it?
5. This article was published in *Science News* magazine. Based on Sumner's word choice, tone, and evidence, who do you think is the article's intended audience? What is the article's explicit or implicit purpose? Use evidence to back up your answers.

Making Connections

6. Research the potential environmental impacts of desalination: Where is desalination already in use? What risks, if any, does it pose to the ocean's ecosystem; how does the technology attempt to avoid or address those

risks; and what is the overall benefit of using this technology in areas where fresh water is limited?

7. Desalination is one of two primary technologies currently being used to address freshwater shortages. The other technology, water reclamation or water recycling, turns wastewater into drinking water. Some major cities, like Las Vegas, already reclaim large portions of their water. Research desalination and water reclamation, then write an essay arguing in favor of one of these methods for a specific region, perhaps your own.
8. Compare the coastal cropping option presented in this article to the vertical indoor farms explained in "The Future of Farming Is Looking Up" (p. 32). Both articles discuss alternative farming options that are especially useful in nontraditional agricultural environments. What factors should regions consider when choosing one of these alternative methods?

Here's How Much It Would Cost to Travel to Mars

Rob Wile and Pascal Lee

Rob Wile has written extensively about both science and economics for a number of publications, including *Money*, *Business Insider*, *Newsweek*, and the *Miami Herald*. In this interview from *Money*, Wile interviews planetary scientist Pascal Lee, cofounder of The Mars Institute and director of the NASA Haughton-Mars Project.

Here, Wile asks not only whether it is physically possible to travel to Mars, but also whether the enterprise would be worth the human and financial cost, and even goes so far as to question what is motivating humans to want to do this to begin with. Lee outlines the many barriers that currently prevent a successful trip, not to mention a colony. Despite these obstacles, many scientists believe that a Mars colony will eventually be a reality and they, along with entrepreneurs like Elon Musk, have already begun mapping out plans to get to Mars and stay there.

It's being billed as the largest event ever dedicated to human exploration to Mars: From May 9 to 11, leading scientists and engineers will gather in Washington for the Humans to Mars Summit.

Among the headline speakers will be Buzz Aldrin°; William H. Gerstenmaier, associate administrator for the Human Exploration and Operations Directorate at NASA; and Pascal Lee, the director of the Mars Institute, an international nonprofit research organization partially funded by NASA.

MONEY spoke with Lee about the challenges facing Mars missions, and why it's important to launch Mars exploration missions despite them.

At This Point, What Would It Cost to Send Someone to Mars?

Pascal Lee: The Apollo lunar landing program cost $24 billion in 1960s dollars over 10 years. That means NASA set aside 4 percent of U.S. GDP to do Apollo. To put things in perspective, we also spent $24 billion per year at the Defense Department during the Vietnam War. So basically, going to the moon with funding spread over 10 years cost the same to run the Department of Defense for one year in wartime.

Now, 50 years, later, today's NASA budget is $19 billion a year; that's only 0.3 percent of GDP, so that's less than 10 times less than what it was in the 1960s. 5

Buzz Aldrin: (b. 1930), American astronaut famous for being the second man, after Neil Armstrong, to walk on the moon.

Meanwhile, the Department of Defense gets $400 billion a year. So the number I find believable, and this is somewhat a matter of opinion, a ballpark figure, doing a human mission to Mars "the government way" could not cost less than $400 billion. And that was going to the moon. This is going to Mars, so you multiply that by a factor of 2 or 3 in terms of complexity, you're talking about $1 trillion, spread over the course of the next 25 years.

As far as sending an "average Jane" to Mars, you're talking even further out in terms of years. Mars is an incredibly lethal environment; there are

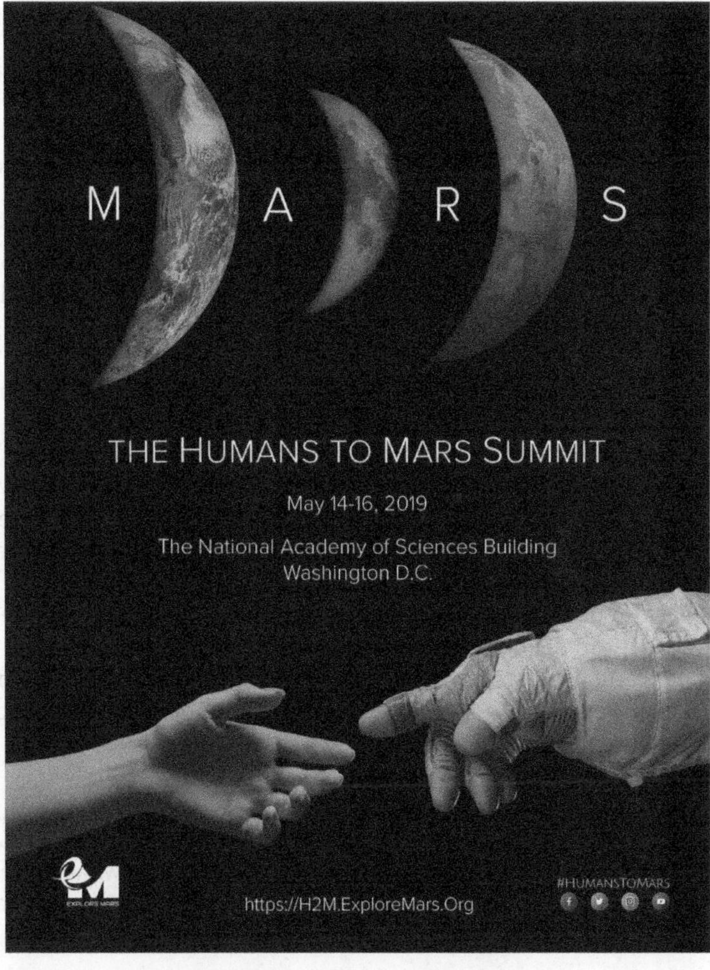

A promotional image for the 2019 Humans to Mars Summit sponsored by Explore Mars Inc., whose goal is to put humans on Mars by 2030.
Explore Mars Inc

several things that can kill you and result in a horrible death if you're exposed unprotected, so, talking about kids growing up on Mars, it's not nearly as soon as you'd hope or think. We instead envision something like Antarctica, where you have maybe a handful of people there at a time performing research for a few decades.

There's another group that wants to turn a Mars mission into, essentially, a reality show, and fund it through commercials, that has other cost estimates, but I find those unrealistic.

What Is the Biggest Cost at This Point?

Pascal Lee: It's not rocketry—the rocket is actually the easiest part, thanks to the private sector. They are able to cut corners, take more risk, do a quick and "not as clean" version. When Elon Musk° presented his vision last October in Mexico, no one questioned his ability to make something big.

The biggest cost is actually to develop all the new systems that would allow us to go to Mars and be productive explorers there. For example, right now, we have a beautiful space suit, but it weighs 300 lbs. On the moon, where gravity is six times less than Earth's, the space suit has a felt-weight of 50 lbs., so that was manageable. But on Mars, the gravity is only about one-third of the Earth's, so you have a space suit with a felt-weight of 125 lbs. That's way too heavy. So there's a technical challenge to cut the mass of the space suit we currently have in half. That's easier said than done.

10

"The reality of Mars is unforgiving, but that's what makes it interesting and challenging and exciting."

We'd also be looking for life, and for that we'll have to develop technology that can dig down beneath Mars' inhospitable surface to find ice or water, and also develop the means of transporting ourselves to locations where that water exists. You're looking at far more complex exploration setup than Apollo.

So What Should Be The Primary Motivation for Going to Mars?

Pascal Lee: The reality of Mars is unforgiving, but that's what makes it interesting and challenging and exciting. It's like President Kennedy said, We would go to Mars not because it is easy, but because it is hard.

Elon Musk: (b. 1971), technology mogul and entrepreneur; CEO of SpaceX and Tesla.

The biggest benefit from the Apollo program at the time was to show the Soviet Union that the Western way of life was better and more capable. The scientific benefit was learning about how the moon was formed and how Earth might have evolved, but the biggest benefit to society was the fact that the Apollo program was a huge shot of steroids into STEM [science, technology, engineering, mathematics] education. The number of grads in science fields and math in mid-60s and early-70s essentially doubled. That became the foundation of the technological competitive edge we have today.

Right now, there's an entire generation of kids in China, enthralled by their space program as they work to send someone to the moon; they're going into STEM fields, and they are basically shaping up to be [leaders] in the next generation in STEM.

Understanding the Text

1. Identify three of the most convincing barriers to actually going to Mars, as explained by Pascal Lee. Are those examples effective in communicating the difficulty of going to Mars?
2. Because this is an interview, there is no clear thesis statement. In the absence of an overtly stated thesis, what do you think is this text's overall argument? How do you know? Identify specific examples to support your answer.

Reflection and Response

3. According to Lee, why would humans actually go to Mars? How is his reasoning rooted in the past but also relevant to the present? Are his reasons compelling enough to convince someone there is a value in going?
4. Because Mars is so far from Earth, much would need to be done to sustain human life on the Red Planet. Lee notes, for example, that "we'll have to develop technology that can dig down beneath Mars' inhospitable surface to find ice or water, and also develop the means of transporting ourselves to locations where that water exists" (par. 11). Visiting Mars will need to be a more invasive process than our trips to the moon have been. What ethical differences are there between visiting the moon and immediately returning to Earth versus going to Mars and staying?

Making Connections

5. Pascal Lee, the interviewee for this article, was a speaker at the 2017 Humans to Mars (H2M) Summit sponsored by Explore Mars Inc. and held in Washington, D.C. Visit the summit's website to learn more about the speakers at this and previous years' events, their presentation topics and materials, and the summit's goals. According to these materials, what do experts think about the probability of traveling to and/or living on Mars?

6. A significant concern for many people across the globe is climate change, which was discussed in "Treading the Fine Line between Climate Talk and Alarmism" (p. 39). As part of a larger discussion on climate change on April 25, 2018, French President Emmanuel Macron addressed the United States Congress and declared that there "is no Planet B." His point was that we only have one Earth and we need to take care of it. After completing some additional research on colonizing Mars as a "Plan B," develop an argument about whether this is a viable plan to preserve humanity if Earth fails.
7. Write an essay comparing the 2015 film *The Martian* (based on the 2011 sci-fi novel by Andy Weir) to the real-world goals, excitement, and fears that people have about going to Mars. Consider how accurately or inaccurately the film portrays Mars itself and the technology necessary to get and stay there. How do these representations contribute to the feelings we already have about traveling to Mars?

The Argument of Future Generations; from *Environmental Ethics Re-Visited*

Francesc Torralba

Francesc Torralba is a philosophy professor at the Universitat Ramon Llull of Barcelona and holds PhD's in both philosophy and theology. As an author of over ninety books, his research and writing focus has been on ethics and human rights. In this excerpt from chapter two of *Environmental Ethics Re-Visited* (2014), which he co-authored with Rosamund Thomas and Miquel Seguró, Torralba argues that the actions we take today will not only affect our descendants but may also determine whether there are future generations. He explains humanity's ethical responsibility by defining and discussing how utilitarianism, third-generation rights, and intergenerational justice should all be studied in an attempt to better understand our environmental legacy. In fact, the idea that current generations owe future generations anything is rooted in history and philosophy, and Torralba's detailed look at the philosophical ideologies of Hans Jonas and John Locke reminds us all that previous generations gave us significant thought. As you read, consider how their perspectives can or should shape our own views.

The Argument of Future Generations

Within the landscape of eco-ethics we need to place the ethics of future generations, which has attracted considerable interest in recent years. From this perspective one must consider not only the value of human life at present, but also the value of human life in the future. This consideration requires undertaking the necessary steps to ensure that such a future exists.

From this perspective the rights of future generations, known as third generation rights, are preserved. These rights do not come from an individualistic tradition as do the first generation rights, or from a socialist tradition as in the case of the second generation rights. Rather, third generation rights derive from a global concern for the future of human life and the perpetuity of the species.

Hans Jonas°, who, to some extent, can be related to this third generation perspective, justifies ethics for future generations in the following terms: "We live in an apocalyptic situation: that is, under the threat of a universal catastrophe if we let things take their present course. About this we have to say something, albeit very well known. The danger comes from the over-dimensioning of the natural-scientific-technical-industrial civilization."[1]

"We must do everything in our power so that our descendants have the means to live better than us."

From the above-stated finding the following imperatives° can be inferred: "An imperative"—Jonas says—that adapts to new kinds of human actions, directed to new kinds of individuals that perform these said actions, would say something like this: "Behave as though the effects of your actions are compatible with the permanence of genuine human life on Earth," or, stated in a negative way: "Act in such a way that the effects of your actions do not destroy the possibility of a genuine and human life," or, simply: "Do not compromise the conditions for an indefinite prolongation of human life on Earth," or even, this time in a positive way: "Include in your present choice, as the object of your will, the future dignity of man."[2]

Closely associated with this idea is the notion of intergenerational justice. Intergenerational justice requires human beings living in the present to oversee the inheritance they have received and not waste it with impunity. From this perspective, we must do everything in our power so that our descendants have the means to live better than us or, at least, that they are not worse off as a result of our actions. 5

The idea of intergenerational justice has been substantiated since Locke's proposal. It can be expressed by the following statements:

a. We need to leave behind an improved land, or, at least, a land that is not impoverished, which requires us not to damage the goods we possess, and improve them by means of our activity.
b. We cannot deplete natural deposits. We have the obligation to leave the same amount of goods that we have received from nature, or at least enough of them for a dignified human development.
c. We need to keep the purity of natural resources intact. We have the obligation to convey this sufficient amount of goods with the same qualities that we received them.

Hans Jonas: (1903–1993), German philosopher concerned with the ethics of modern technology.
Imperative: A command.

Hans Jonas holds these views but from the standpoint of his ethics of responsibility. Jonas reverses the utilitarian argument on the ethics of responsibility towards future generations. While utilitarianism° maintains that the priority must be to maximize welfare (pleasure, practical use) and avoid suffering in order to justify a human life worth living, Jonas works on the assumption that the first obligation is towards the existence of humanity, and from here many duties derive, aimed at establishing the conditions which allow this defense of the essence of humanity.

Jonas provides two axioms° for the foundation of a future-oriented ethics: first, there must be a future, and second, it is unlawful to bet on the existence of humanity as a whole. From the first axiom we derive two duties: to ensure an adequate representation of our future, and an appeal to a sense of fear (the heuristics of fear). The obligations towards future humanity are based on freedom and a dignified life. According to Jonas, we have to ensure the realisation of an authentic humanity.

Guided by Max Weber's° ethics of conviction and of responsibility, Jonas suggests the integration of future generations in moral judgments. To this end, he redefined the Kantian principles:°

a. Acting so that the effects of one's actions are compatible with the permanence of a genuine life on this planet.
b. Acting in such a way that the effects of one's actions cannot damage the expectations of human life in the future.
c. Not jeopardizing the conditions for an indefinite continuity of human life on Earth.

In summary, while utilitarianism makes the foundation of existence depend on whether certain conditions of life are achieved, the deontological position of Jonas reverses these terms: our first duty must be to focus on the existence of a future humanity and from here we derive the absolute obligation to ensure conditions that make life possible, so that ultimately human essence reaches the standards of decent life.

One of the most important ethicists of future generations is Professor Giuliano Pontara (b. 1932). This author looks closely at the symptoms of the ecological crisis to justify properly the need for an ethics of future

Utilitarianism: philosophy wherein the morally right act is the one that benefits the most people.
Axioms: statements regarded as being generally true by the majority of people.
Max Weber: (1864–1920), German sociologist and economist.
Kantian principles: moral imperatives developed by German philosopher Immanuel Kant (1724–1804).

generations. He argues that energy sources are at risk of running out and that the impoverishment, or depletion, of fossil and nuclear resources can have far-reaching negative consequences at different levels for those who will inhabit the planet in two or three hundred years. He also notes that fossil and nuclear fuels are highly polluting and already have caused severe damage.

Pontara shows that the greenhouse effect can have serious consequences for large populations of future generations given that it tends to produce a global warming. And consequently, on the one hand, a desertification of areas that are currently covered with vegetation, and, on the other hand, a partial loss of glaciers, a significant rise in sea levels and, therefore, the flooding of vast territories which are currently densely populated.

He also focuses on the problem created by so-called holes in the ozone layer which, according to Pontara, is another dramatic consequence that might affect future generations. The Earth is surrounded by an ozone layer, the ozonosphere, which protects humans, animals and plants from excessive doses of ultraviolet radiation. After the introduction of carbon chlorofluorocarbons from different types of processes and industrial products, the ozone in the ozonosphere is attacked by chlorine and transformed into oxygen. Consequently, the thickness of the ozone layer tends to weaken and, therefore, it filters a lesser amount of ultraviolet rays, which increases the risk of skin cancer and other dangerous diseases for man and even animals and plants. Pontara shows the increasingly worrying dangers associated with nuclear fuels, for both present and future generations. He also mentions the threat of radioactive waste—an additional problem related to nuclear fuels. The main dangers for future generations coming from radioactive waste are those relating to the contamination of soil, water layers, oceans, polar regions, and by extension, outer space. Ultimately, it will all depend on whether most hazardous waste is placed deep in the Earth's crust or at the bottom of the ocean, in areas considered to be geologically safe, or, if it is stored in big containers in outer space.

A further factor that might cause serious problems to future generations, as Pontara puts it, is the process of increasing pollution and the impoverishment of fresh water reserves. The contamination process of deep water layers can be particularly harmful for future generations. This despoiling might take place after the invasion of salt water caused by rising sea levels due to global warming or, as already indicated, by the pollution of radioactive waste if placed deeply underground. Once this process of contamination has begun it might be unstoppable, or reversible only at very high costs.

Other threats for future generations are, according to Pontara, those related to the processes of desertification and pollution of arable land for the following reasons: erosion; population pressure; neglect; inadequate drainage; inadequate irrigation facilities; the use of pesticides and chemical fertilisers; and others. Among the various possibilities we currently have to influence life for future generations we must mention those that have appeared due to the development of biomedical science, biotechnologies, and genetic engineering. Each of these opens the possibility of eugenics: that is, programming the existence of human beings with certain qualities and of certain types, conditioning those that constitute the future generations themselves. The key question here is the following: must we influence? In case of the answer being affirmative: in what way? 15

Notes

1. H. Jonas, *El principio de responsabilidad,* Herder, Barcelona, 1995, p. 233.
2. Ibid., p. 40.

Understanding the Text

1. Torralba references and explains multiple philosophical perspectives. What are the differences and connections between third-generation rights, intergenerational justice, future-oriented ethics, and utilitarianism?
2. What is the main argument that Torralba is making in this piece? Choose two or three pieces of evidence to support your answer.

Reflection and Response

3. From an eco-ethics standpoint and the related intergenerational justice perspective, Torralba argues that "one must consider not only the value of human life at present, but also the value of human life in the future" (par. 1). Do you agree with this statement? Why or why not?
4. *Environmental Ethics Re-Visited* is a book largely made up of writing on scientific methods, but this excerpted essay is more philosophical than scientific in its approach. What purpose do you think a reading about the ethical choices in science serves?
5. The title of this excerpt is "The Argument of Future Generations." Why do you think Torralba chose the word *of* instead of *for*? How would *for* change the title's meaning?

Making Connections

6. Investigate the connections between utilitarianism and environmentalism. Based on your research, examine why these two perspectives could easily complement or contradict one another.

7. Filmmakers have long examined what future generations will do once the Earth's resources have been depleted. From children's films like *WALL-E*, where humans have had to abandon a garbage-filled Earth, to science-fiction movies such as *Avatar*, where the human race is systematically taking over the ecosystem of the native peoples to mine it for its own uses, audiences are interested in where we might go next. Choose two stories or films that deal with a dystopian Earth and where natural resources are scarce. What behaviors are being exhibited by the humans? Do you see any direct connections between them and the ethical philosophies presented here?

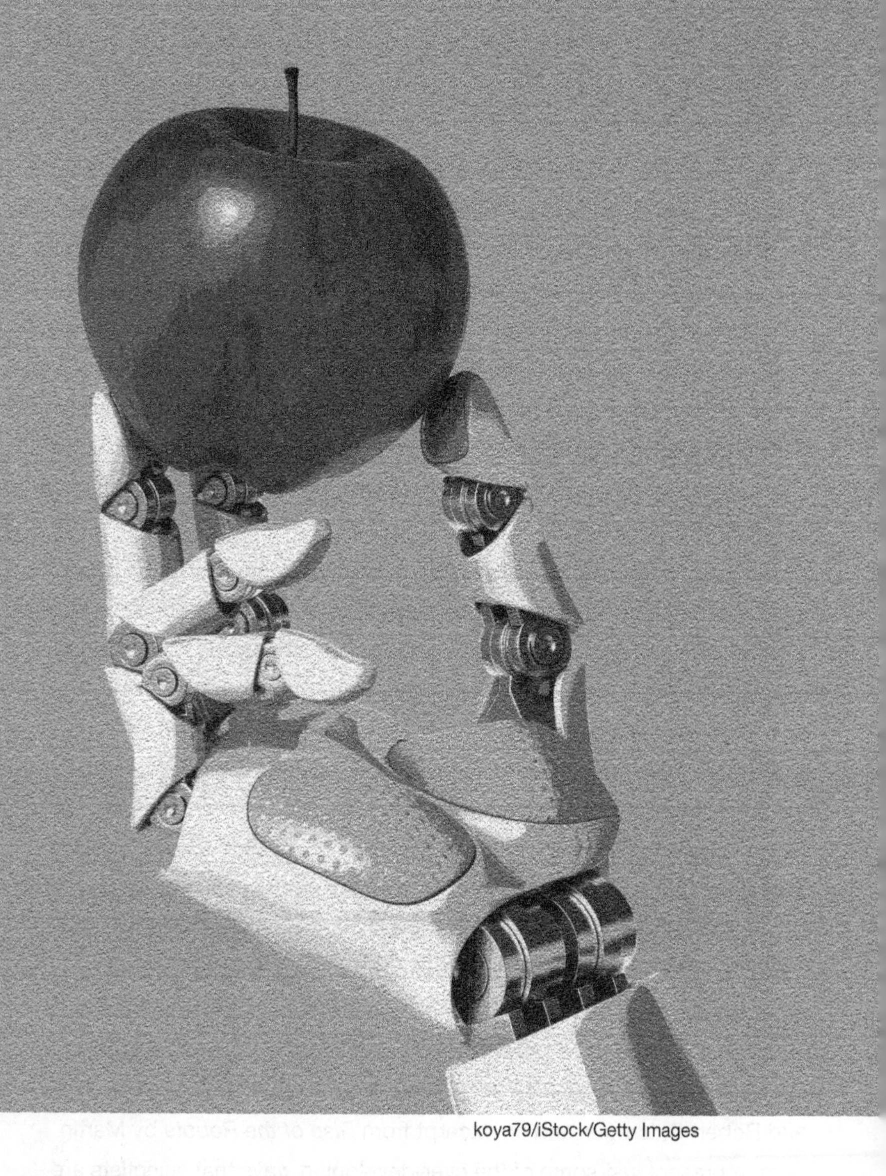
koya79/iStock/Getty Images

2 Is Technology Surpassing Biology?

In his book *Life 3.0: Being Human in the Age of Artificial Intelligence* (2017), Max Tegmark, a physicist and cosmologist at MIT, argues that human life is progressing in three stages that he identifies as 1.0, 2.0, and 3.0. In the third and final stage, humans are able to design hardware and software that can be uploaded to their bodies, and humans become free of their "evolutionary shackles." While Tegmark notes that humans have not yet progressed to this stage, he methodically examines the opinions of other prominent scientists on two things: when we will reach stage 3.0, and what it will mean when we do. He calls this "the most important conversation of our time." But why? Are we really that close to 3.0?

Science and technology have advanced significantly just in the last thirty years: the human genome is mapped; DNA can be replicated and synthesized; and genetic engineering of human embryos is now possible. Humans have also become increasingly comfortable with the mixing of science and technology with their own biology. Pacemakers and prosthetic limbs have advanced toward near-seamless integration into the lives of those who depend on them. These are just two examples of ways humans rely on technology to perform biological functions and stay alive, or maintain a high quality of life. Not as common yet, but already in use, are wearable exoskeletons that will increase human strength and mobility, and microchips holding all of our personal information that can be implanted under our skin. The technologies highlighted in the first four readings of this chapter, "New Natural Selection" by Bryan Walsh, "The CRISPR Pioneers" by Alice Park, "Rebuilding Ourselves" by Rohit Karnik and Robert S. Langer, and an excerpt from *Rise of the Robots* by Martin Ford, represent just some of the ever-developing ways that scientists are looking to more effectively diagnose, treat, and/or eliminate disease. While these technologies certainly have the potential to improve the quality of human life or significantly extend it, some do so by co-opting natural biological processes, and all will translate into drastic shifts in traditional medical treatment.

photo: koya79/iStock/Getty Images

As worthwhile as some of the goals detailed in many of these readings may sound, others, such as Hillary Rosner's "All Too Human" and Jon Cohen's "The Horror Story That Haunts Science," dig more deeply into philosophical and ethical questions that naturally develop with any new science. These readings examine the human desires that motivate these technologies, and consider the moral and ethical implications of "fixing" naturally occurring problems, like disease and aging. Current developments in science and technology demand that we question whether we want to create, like Frankenstein did, superhumans that are impervious to disease. If we accomplish this, would it, then, be acceptable to go further and genetically engineer the human body to be disease resistant? "All Too Human," and Gary Shteyngart's "Super Sad True Love Story," a fictional piece, both ask us to also consider the emotional and psychological effects of human enhancement and whether or not we really understand what we would get if we pursued our scientific desires to their most extreme ends.

Ultimately, science now has the ability to rewrite and replace the biology of a human being, so more human beings should be conscious of how and why this is occurring. The readings in this chapter help to ensure that all stakeholders, not just scientists, are part of these conversations.

New Natural Selection: How Scientists Are Altering DNA to Genetically Engineer New Forms of Life

Bryan Walsh

Bryan Walsh is a freelance science, public health, and environmental writer. Although he worked for over fifteen years at *Time* magazine, he has also been published in other magazines like *Newsweek*, and he is currently working on *End Times*, a book about the end of the world.

In this cover story for *Newsweek*, he discusses what he calls the limits of the natural world and examines the science of synthetic biology, which enables scientists to "edit genomes and even write entirely original DNA" (par. 4). Walsh explains the process of synthetic biology as offering another way to understand the natural world and the biological processes of individual organisms. What we may assume is a simple organism, like bacteria, is often far more complex than first thought, and by trying to recreate it, scientists actually learn more about its make-up. Synthetic biology is also a booming business, and the race to perfect it is not only scientifically, but financially, worthwhile. With all of its vast potential for good, however, Walsh's piece reminds us that scientists must consider its potential risks.

Before human beings wrote books or did math or composed music, we made leather. There is evidence hunter-gatherers were wearing clothes crafted from animal skins hundreds of thousands of years ago, while in 2010 archaeologists digging in Armenia found what they believed to be the world's oldest leather shoe, dating back to 3,500 B.C. (It was about a women's size 7.) For a species sadly bereft of protective fur, being able to turn the skin of cows or sheep or pigs into clothing with the help of curing and tanning would have been a lifesaving advance, just like other vital discoveries Homo sapiens made over the course of history: the development of grain crops like wheat, the domestication of food animals like chickens, even the all-important art of fermentation. In each case, human beings took something raw from the natural world—a plant, an animal, a microbe—and with the ingenuity that has enabled us to dominate this planet, turned it into a product.

The natural world has its limits, though. Tanned animal skin may make for stylish boots, motorcycle jackets and handbags—supporting an industry worth about $200 billion a year—but it's still animal skin. That would seem to be an insurmountable problem if you're one of the hundreds of millions of vegetarians around the world, or even just someone who worries about the environmental impact of raising tens of billions of animals for clothing and food. But it's not the animal skin that makes leather leather—it's collagen, a tough, fibrous protein that is a major biological component of animal connective tissue, including skin. If there was a way to manufacture collagen alone, it might be possible to produce leather that even the most dedicated animal-rights activist could love.

And that's exactly what's happening on the eighth floor of the cavernous Brooklyn Army Terminal on New York's waterfront, where Modern Meadow has its labs and offices. There, the 60-person startup takes tiny microbes and edits their DNA—the genetic code that programs their behavior—so they will yield collagen as a metabolic product, just as the yeast that brew beer create alcohol from grain sugar. The result is a microbiological factory, as the tweaked cells multiply in vats and the harvested collagen is processed. After a tanning procedure—one more sustainable than that used in standard tanning, since there is no animal hair or fat to remove from the microbe-grown collagen—what's left is a material that is biologically and chemically similar to conventional leather, save chiefly for the fact that no animals were harmed in its making. In fact, this biofabricated leather may be better than animal leather—Modern Meadow's microbes can produce collagen much faster than it would take to raise a cow or sheep from birth, and the company can work with brands to design entirely new materials from the cell level up. "It's biology meets engineering," says Andras Forgacs, the cofounder and CEO of Modern Meadow. "We diverge from what nature does, and we can design it and engineer it to be anything we want."

> "We diverge from what nature does, and we can design it and engineer it to be anything we want."

That is the promise of synthetic biology, a technology that is poised to change how we feed ourselves, clothe ourselves, fuel ourselves—and possibly even change our very selves. While scientists have for decades been able to practice basic genetic engineering—knocking out a gene or moving one between species—and more recently have learned to rapidly read and sequence genes, now researchers can edit genomes° and even write entirely original DNA. That gives scientists incredible control over the fundamental code that drives all life on Earth, from the most basic bacterium to, well, us. "Genetic engineering was like replacing a red light bulb with a green light bulb," says James Collins, a biological engineer at the Massachusetts Institute of Technology and one of synthetic biology's early pioneers. "Synthetic biology is introducing novel circuitry that can control how the bulbs turn off and on."

Genome: a complete set of genes in a cell or organism.

We can use that control to harness nature to our own ends and do 5
so in a way that will help solve some of our most pressing sustainability challenges. Cells could be engineered to make meat in a lab, eliminating the need for environmentally intensive and often cruel factory farms. Bacteria could be manipulated to secrete oil, providing a truly renewable source of liquid fuel. Yeast could be designed to produce artemisinin, a vital antimalarial drug that in its natural form must be made from limited supplies of the sweet wormwood plant—which, as it happens, is already being done. "What is at stake here is finding a way to make everything humans need without trashing our civilization," says Drew Endy, a synthetic biologist at Stanford University who helped launch the field. "We can transition from living on Earth to living with Earth."

Market-Driven Eugenics

The dawn of the Synthetic Age is not just the province of scientist-dreamers and Brooklyn startups. A 2016 Transparency Market Research report predicted that the synthetic biology market would grow from $1.8 billion in 2012 to $13.4 billion by 2019. Last year, synthetic biology companies took in $1 billion from investors—including tech titans like Eric Schmidt, Peter Thiel and Marc Andreessen—double the total from 2014. Even the giants of the fossil-fuel world have gotten in on the game—Exxon Mobil has a $600 million deal with Synthetic Genomics, a partnership that bore fruit in June when the company announced a major breakthrough in engineering strains of algae to produce oil for use in sustainable biofuels.

The true benefits—and consequences—of synthetic biology will come as scientists move from mimicking nature in the lab to redesigning it. Imagine plants that change color in the presence of explosives or microbes that can secrete the scent of a long-extinct flower. Picture a cell line that is immune to all bacteria and viruses, or even the 3 billion DNA base pairs of a human being's genome, fully synthesized in a lab. All of those projects are underway at various stages, and the last target—writing an entire human genome—would be an epochal achievement for science, potentially opening the door to re-engineering the human body itself, making us healthier, smarter, stronger. It's one of the goals of GP-write, an international project launched in 2016 by a group of synthetic biologists who want to stimulate the development over the next decade of technology that could synthesize the genomes of large organisms—including humans. "Being able to write large genomes means moving from natural selection and artificial selection—think traditional plant and animal breeding—to intentional design," says Andrew Hessel, distinguished research scientist at the design company Autodesk and one of the founders of GP-write.

If the idea of synthesizing an entire human genome alarms you, you're not alone—even some synthetic biologists, like Stanford's Endy, are wary of the notion. The researchers behind GP-write have made it clear that they have no intention of creating artificial people with their synthesized DNA; rather, their work will be confined to synthesizing human cells, in an effort to better understand how the human genome works—and, potentially, how to make it work better. But any attempt to engineer the genetic code of living beings raises ethical concerns—first over safety, and even more so, over success. What happens if an engineered plant or animal escapes into the wild, where its impact on the environment would be hard to predict? Engineering human cells to eliminate deadly genetic disorders might seem straightforward, but where would we draw the line between treatment and enhancement? "We're developing powerful tools that are changing what it is to be human," says Jim Thomas, a researcher with the technology watchdog the ETC Group. "The worry is that you could have market-driven eugenics."

Of course, those ethical questions assume that synthetic biologists will be able to replicate a human genome—and that's far from certain. Scientists have yet to fully synthesize the genomes of much simpler single-celled organisms like yeast, so it could take far longer than a decade to learn how to write the 20,000 or so genes in a human genome. And like all technologies moving from the lab to the real world, synthetic biology will need to compete with conventional products in the market and at scale. Over the past several decades, startups that employed the tools of synthetic biology to produce advanced biofuels burned through hundreds of millions of dollars in a mostly futile effort to beat cheap gasoline. But whether it happens in the near term or the long term, the science behind synthetic biology—the ability to read and write the code of life—is already with us. And it's poised to reengineer the world as we know it.

Synthetic Biology

Look out your window. Every bit of living matter you see—the tree bending toward the sun, the sparrow winging on the breeze, the person walking past—operates on the same genetic code, the nucleobases of DNA: cytosine (C), guanine (G), adenine (A), thymine (T). This is the programming language of life, and in its basics it hasn't changed much since it arose out of Earth's primordial ooze. Just as the English language can be used to write both "Baa, Baa Black Sheep" and *Ulysses* so can DNA in all its

10

combinations write the genome of a 2 mm long E. coli bacterium and a 30-meter-long blue whale. "The same DNA in humans is the same DNA in every organism on the planet," says Jason Kelly, the CEO of Ginkgo Bioworks, a synthetic biology startup based in Boston. "This is the fundamental insight of synthetic biology."

The language of DNA may have been written billions of years ago, but we learned to read it only in recent years. Sequencing DNA—determining the precise order of the C, G, A and T—was first done only in the 1970s, and for years it was laborious and expensive. It took more than 10 years and about $2.7 billion for the scientists behind the Human Genome Project to complete their mission: the first full sequence draft of the genes that encode a human being. But thanks in part to technology advances driven by that public-private effort, the price of sequencing DNA has plummeted—it now costs around $1,000 to sequence a person's full genome—even as the speed has multiplied, to little more than a day.

If that sounds familiar, it should—the same thing happened to the cost and speed of microchips over decades past, as Intel cofounder Gordon Moore predicted in the law that bears his name. And just as faster and cheaper microchips drove the computer revolution from the days of room-size mainframes through the dawn of the iPhone, so does cheap DNA reading—and increasingly, writing—make possible the revolution of synthetic biology. "It wasn't merely that you could do it, but that the costs came down so much starting 15 to 20 years ago," says Rob Carlson, the managing director of Bioeconomy Capital. "The improvements were even faster than Moore's Law."

Carlson should know. The Carlson Curve, the biotech equivalent of Moore's Law, was named after him—though like many in the field, he doesn't care for the name. (Carlson preferred "intentional biology," but biologists demurred—they thought the term made it sound as if their work hadn't been intentional before.) Synthetic denotes fake and artificial, the imitation of natural—think synthetic fabrics like nylon and polyester—but that's not how most synthetic biologists view their craft. To Stanford's Endy, who helped launch the movement when he was at MIT more than a decade ago, synthetic biology is about understanding the messy process of life at the cellular level and above through the process of engineering it. "We don't know how to do it, so we try, and one learns by doing," says Endy.

There's a quote that synthetic biologists repeat like a mantra, from the great theoretical physicist Richard Feynman: "What I cannot create, I do not understand." Feynman didn't exactly say those words—the phrase was found on the physicist's blackboard at CalTech at the time of his death.

But to synthetic biologists, it means that the process of editing and writing DNA—engineering life—is necessary for us to better understand how DNA works. To that end, scientists have worked to synthesize genomes—meaning writing and printing whole artificial genes, rather than copying existing DNA, as in cloning—of organisms, starting with the simplest ones, in an effort to understand what the words in the genetic book of life really mean. An initial success came in 2010, when the geneticist Craig Venter—who helped lead the Human Genome Project—and his colleagues created the first synthetic cell, writing the entire genome of a tiny bacterium called Mycoplasma mycoides and inserting it into the empty cell of another bacterium. (They nicknamed the cell Synthia.) That was a remarkable achievement in its own right, but in 2016 Venter and his team went one better, taking Synthia's genome and methodically breaking it down until they reached the minimum number of genes required to sustain life. By stripping life to its basics, researchers could discover what each gene actually did. "The aim is to make something simple, to remove complexity, so you can begin engineering," says Sophia Roosth, a historian of science at Harvard University and the author of the new book *Synthetic: How Life Got Made.*

15 As it turned out, even the world's simplest bacterial genome was more complicated than scientists might have suspected. Of the 473 genes in Venter's pared-down, synthetic cell, the functions of 149 were completely unknown. That's almost a third, which underscores how far scientists have to go before they can truly claim to understand the genetic code that we can now sequence so easily—let alone effectively synthesize the genomes of much bigger and more complex organisms. It brings to mind another Feynman quote—"the difference between knowing the name of something and knowing something."

Rewriting the Book of Life

To actually know, scientists will need to sequence, synthesize, and program vast amounts of genetic data. You design an organism, build it—through DNA synthesis or through gene-editing tools like CRISPR—test it out in the lab and, hopefully, learn from the experiment. Then you do it again, and again, in a cycle called design-build-learn-test. In the case of Venter's synthetic cell, for instance, scientists would add or subtract a gene at a time and then look to see what happened to their organism. If the synthetic bacterium died, that was a pretty good sign that the gene in question was important. "That to me is the heart of engineering biology," says Nancy Kelley, another of the founders of GP-write.

But you'll only be able to learn as fast as you can design and build and test. That's why Carlson's Curve is so important. Think computer programming, which advances on a similar cycle. When Tom Knight, one of Jason Kelly's cofounders at Ginkgo Bioworks, was helping to build what would become the internet at MIT in the 1960s, he was programming on refrigerator-sized computers that required users to manually enter deck after deck of punched cards. It was slow and laborious—and that was about the speed of biological programming until fairly recently. "We would spend an entire afternoon doing by hand site-directed mutagenesis that would enable you to change a single A to a T in the genome of a bacteria," says Kelly. "That's like spending an afternoon changing a bit from a zero to a one on a computer."

Today, as Kelly notes, "a guy at Facebook can create a new product in a single afternoon," simply because computers have gotten so much faster. We may never be able to program biology as fast as we can a computer—in part because biology is made up of matter, however tiny, whereas computer code is just code—but we will keep getting faster. "Back in 2002 and 2003, it used to cost me $4 to press the DNA synthesis button once for one letter," says Endy. Now, says Emily Leproust, the CEO of San Francisco–based DNA synthesis startup Twist Bioscience, her company can synthesize a base pair—the essential building blocks of the DNA helix—for just 9 cents.

In the design-build-test-learn cycle of synthetic biology, Twist supplies the building materials. Labs and companies send orders for specific genes to Twist, and the company does the work of synthesizing them, printing tiny molecules of DNA on silicon. The turnaround time is a matter of weeks, and as Twist and other DNA synthesis companies get better, that will be shortened. As the barriers of cost and time fall, what is liberated is the imagination of synthetic biologists, who are able to rapidly try out ideas. Just as the dawn of the internet led to an array of tech startups in the 1990s—some of which are now pillars of the global economy, like Amazon and Google—so the commercialization of DNA-writing technology is giving birth to a fresh industry. Apple founder and CEO Steve Jobs, shortly before his death from cancer, told his biographer that "I think the biggest innovations of the twenty-first century will be at the intersection of biology and technology. A new era is beginning."

Many companies and investors are convinced that synthetic biology could revolutionize some of the fundamental ways we live and do business. Technologists at companies including Microsoft believe that DNA could even overtake silicon in our hard drives as a storage medium. The genetic code, after all, is really just a means to preserve and transmit information—the information of how a living thing works. DNA is an incredibly dense medium—researchers this year developed a method theoretically capable of storing all of the data in the world on a single room's worth of DNA—and unlike existing physical recording mediums, there's no danger of it being made obsolete. Biology, after all, has been writing DNA for billions of years.

Twist has begun working with Microsoft to perfect the process of DNA storage, and in April the software company purchased 10 million strands of DNA from Twist as part of that agreement. "As much as the last century was about plastics, this century will be about biology," says Leproust.

At Ginkgo Bioworks, the biggest consumer of synthetic DNA on the planet, they can smell the future coming. Founded in 2008 by Kelly and four of his colleagues from MIT's pioneering synthetic biology program, Ginkgo designs customized living organisms—engineered baker's yeast—that can produce flavors and fragrances that are usually derived from plants. Ginkgo has partnered with French perfume company Robertet to create a rose fragrance by extracting the genes from real roses, injecting them into yeast and then engineering the microbe's biosynthetic pathways to produce the smell of a rose—which apparently smells just as sweet when emitted from a yeast. That might come as a surprise to some consumers who don't know that the active ingredient in their perfume came from engineered microbes. But it's worth noting that the yeast itself isn't a part of the perfume, and the rose oil it produces has a claim to be far more natural than any chemical substitute. "We said, What if instead of going into a field of roses to get rose oil, you can run a brewery?" says Kelly. "And instead of brewing beer, you brew rose oil? We develop those designed yeast using our platform, and we license it out to our customers."

Synthetic biologists won't be satisfied with simply copying existing forms of life; they want to engineer something new, and even bring long-dead organisms back to life. Ginkgo is working on extracting DNA molecules from plant specimens preserved in herbariums, to synthesize the fragrances of flowers that have gone extinct, like an olive bush from the South Atlantic island of St. Helena that disappeared from the wild in 1994. The Bay Area startup Bolt Threads has engineered yeast microbes that can secrete spider silk, a material that is stronger than steel and yet extremely lightweight. Bolt has already used the spider silk thread to make ties, but the superstrong material could have a future in pharmaceutical products and the military. (The company is also an excellent example of why the decrease in the cost of DNA synthesis is so important—it has gone through some 4,000 formulations to properly engineer yeast capable of making spider silk.) In the lab, Colorado State University biologist June Medford is working with the Defense Department (and its $7.9 million grant) to engineer plants that would turn white in the presence of a bomb. Medford imagines that the engineered plants—which are likely years away—could be used in airport security lines, perhaps in place of multimillion-dollar wave scanners. "Plants have developed over 4 billion years to sense and respond to their environment," she says. "We identify a synthetic biology component that enables that and plug it into the natural infrastructure."

Infrastructure is an apt term. Right now, ours is powered mostly through minerals and petrochemicals, but synthetic biology offers the possibility of an infrastructure with built-in sustainability. As a farmer's field reliably demonstrates every spring, biology is renewable in a way that coal or oil or iron simply isn't. Biology is also simply very, very good at what it does, which is sustainable growth. All the plants on Earth, Endy says, harness 90 terawatts of energy, which he notes is about four and a half times the energy currently used by humanity. A biological cell can carry out complex operations far beyond the scope of our smartest artificial intelligence. "Biology is better at making small precise things than Intel is, and it makes more big physical stuff than car companies—all in a sustainable way," says Kelly. A pine tree, for example, is infinitely more complex and has a longer life than a Lexus.

25 Looking into the future, synthetic biologists think they may be able to program cells to grow into almost anything. "Imagine your iPhone being grown from an engineered design, with cases made from synthetic leather and a screen that produces its own light," says John Cumbers, the founder of SynBioBeta and coauthor of the forthcoming book *What's Your Bio Strategy?*

It's going to be a long time before we're harvesting iPhones in the fields. As cheap and as fast as DNA synthesis has gotten, it needs to be much cheaper and much faster. It may cost less than a dime to synthesize a single DNA base pair now, but Kelly points out that if a tech company like Facebook had to spend even a penny every time it changed a single bit in a software program, there would be no money left for anything else. "We're still in the IBM era of this technology," he says. DNA synthesis companies are limited in the length of DNA strands they can produce at a time—Twist's maximum, for instance, is about 3,200 base pairs long. To put that in perspective, the entire human genome is roughly 3 billion base pairs long. That means researchers need to take those strands and link them together—not impossible, but hardly seamless either. "There's a messiness to biology, and engineering it is hard," says MIT's Collins. "It's a lot harder than we thought it would be."

Understanding the Text

1. The term *synthetic biology* is used frequently in this article. Based on what you read in the article, what do you think this term means?
2. Walsh explores numerous potential uses of synthetic biology. Identify three of them – what are the goals of each one and what is their current stage of development?

Reflection and Response

3. This article was originally published in *Newsweek*, a magazine with a general audience, not necessarily one with a scientific background. Consider the language used in this article and the concepts being explained. Do you think Walsh explained these concepts effectively for a general audience? Using examples, explain your answer.
4. Walsh notes that "cells could be engineered to make meat in a lab, eliminating the need for environmentally intensive and often cruel factory farms" (par. 5). How would this affect you? Would you eat meat grown in a lab rather than naturally farmed? Explain your reasoning.

Making Connections

5. Do some research on CRISPR, in addition to reading "The CRISPR Pioneers" (p. 82). What is CRISPR and what can it do? What are its potential benefits and who do they serve? What are its known risks? Considering all of this, are there "compelling reasons" for moving forward with this technology?
6. Visit the Synthetic Biology Project through the Woodrow Wilson International Center for Scholars website and explore some of the most recent applications of synthetic biology. What trends do you see? What surprises you about your findings?
7. Choose a film that centers on genetic engineering or human modification and examine some of the motivations behind the fictional science. Do these motivations apply to real life? What are the possible benefits and risks, as seen in the film? Could those apply to real life as well?

The CRISPR Pioneers

Alice Park

Alice Park has been a health and medicine reporter for *Time* for over twenty-five years. She is a two-time CASE Media fellow and authored the book *The Stem Cell Hope: How Stem Cell Medicine Can Change Our Lives*. In this *Time* magazine cover story, Park introduces readers to the scientists behind the revolutionary gene-editing technology known as CRISPR°, which has the potential to, among other things, turn immune cells into cancer-fighting cells.

According to the American Cancer Society data, in 2018 alone, there were over 1.7 million new cancer diagnoses. The human immune system is often powerless to fight off cancer on its own, so the hope that CRISPR can provide is significant. However, CRISPR's many potential uses are also cause for significant and valid concern. Park points out that "CRISPR allows scientists to easily and inexpensively find and alter virtually any piece of DNA in any species" (par. 3) and this has already been done with some foods, animals, and as recently as 2018, even human babies.

Dr. Carl June's lab at the University of Pennsylvania looks like any other biology research hub. There are tidy rows of black-topped workbenches flanked by shelves bearing boxes of pipettes and test tubes. There's ad hoc signage marking the different workstations. And there are post-docs° buzzing around, calibrating scales, checking incubators and smearing solutions and samples onto small glass slides.

Appearances aside, what June is attempting to do here, on the eighth floor of the glass-encased Smilow Center for Translational Research in Philadelphia, is anything but ordinary. He's built a career trying to improve the odds for people with intractable end-stage disease, and now, in the university's brand-new cell-processing lab, he's preparing to launch his most ambitious study yet: he's going to try to treat 18 people with stubborn cancers, and he's going to do it using CRISPR, the most controversial new tool in medicine.

Developed just four short years ago by two groups—Jennifer Doudna, a molecular and cell biologist at the University of California, Berkeley, together with Emmanuelle Charpentier, now at the Max Planck Institute in Berlin; and Feng Zhang, a biomedical engineer at the Broad Institute of Harvard and MIT—CRISPR allows scientists to easily and inexpensively

CRISPR: Clustered Regularly Interspaced Short Palindromic Repeats.
Post-doc: someone who has a doctoral degree and is pursuing temporary professional mentorship to gain professional skills in a specific field.

find and alter virtually any piece of DNA in any species. In 2016 alone it was used to edit the genes of vegetables, sheep, mosquitoes and all kinds of cell samples in labs. Now, even as some scientists call for patience and extreme caution, there's a worldwide race to push the limits of CRISPR's capabilities.

June's ultimate goal is to test CRISPR's greatest potential: its ability to treat diseases in humans. "Before we were kind of flying in the dark when we were making gene changes," he says of earlier attempts at genetic tinkering. "With CRISPR, I came to the conclusion that this technology needs to be tested in humans." The trial, which will start treating patients in a few months, is the first to use this powerful technique in this way. It represents the most extensive manipulation of the human genome ever attempted.

5 Soon, June's 18 trial patients will become the first people in the world to be treated with CRISPR'd cells—in this case, cells genetically edited to fight cancer. Like many people with cancer, the patients have run out of options. So, building on work by Doudna, Charpentier and Zhang, June's team will extract their T cells, a kind of immune cell, and use CRISPR to alter three genes in those cells, essentially transforming them into superfighters. The patients will then be reinfused with the cancer-fighting T cells to see if they do what they're supposed to do: seek and destroy cancerous tumors.

A lot of hope hangs on the outcome of the trial, but whether it succeeds or fails, it will provide scientists with critical information about what can go right and wrong when they try to rewrite the genetic code in humans. The hope is that studies like June's will bear out CRISPR's therapeutic potential, leading to the development of radical new therapies not just for people with the cancers being studied but for all of them, as well as for genetic diseases such as sickle-cell anemia and cystic fibrosis, and chronic conditions like Type 2 diabetes and Alzheimer's. It may sound far-fetched, but studies like this one are an enormous first step in that direction.

Using CRISPR on humans is still hugely controversial, in part because it's so easy. The fact that it allows scientists to efficiently edit any gene—for some cancers, but also potentially for a predisposition for red hair, for being overweight, for being good at math—worries ethicists because of what could happen if it gets into the wrong hands. As of now, the National Institutes of Health (NIH), by far the world's largest sponsor of scientific research, will not fund studies using CRISPR on human embryos. And any new way of altering genes in human cells must get ethics and safety approval by the NIH, regardless of who is paying for it. (The NIH also opposes the use of CRISPR on so-called germ-line

Silicon Valley is not new to the DNA business. One of its most popular startups is 23andMe, the at-home kit for DNA testing.
Martin Shields/Alamy Stock Photo

cells—those in an egg, sperm or embryo—since any such changes would be permanent and heritable.)

To fund his study, June was able to attract support from Sean Parker, the former Facebook executive and Silicon Valley entrepreneur behind Napster. Parker recently founded the $250 million Parker Institute for Cancer Immunotherapy°, a collaboration among six major cancer centers, and June's study is its first ambitious undertaking. "We need to take big, ambitious bets to advance cancer treatment," says Parker. "We're trying to lead the way in doing more aggressive, cutting-edge stuff that couldn't get funded if we weren't around."

That's not to say June's study will necessarily cure these cancers. "Either it's back to the drawing board," he says, "or everyone goes forward and studies a wide variety of other diseases that could potentially be fixed." In reality, both things are probably true.

10 Even if June's study doesn't work as he hopes, experts still agree it will be a matter of months—not years—before other privately funded human studies get launched in the U.S. and abroad. An ongoing patent

Immunotherapy: a type of cancer treatment that assists the immune system in fighting cancer naturally.

battle over who owns the lucrative technology hasn't stopped investors from pouring millions into CRISPR companies. So simple and inexpensive is the technique, and so frenzied is the medical community about its potential, that it would be foolish to bet on anything else. "With a technology like CRISPR," says Doudna, "you've lit a fire."

A Year of Progress

CRISPR's journey from lab bench to cancer treatment may seem quick. After all, as recently as a couple of years ago only a minuscule number of people even knew what clustered regularly interspaced short palindromic repeats—that's longhand for CRISPR—was. But the technology is at least hundreds of millions of years old. It was bacteria that originally used CRISPR, as a survival mechanism to fend off infection by viruses. The ultimate freeloaders, viruses never bothered developing their own reproductive system, preferring instead to insert their genetic material into that of other cells—including bacteria. Bacteria fought back, holding on to snippets of a virus' genes when they were infected. The bacteria would then surround these viral DNA fragments with a genetic sequence that effectively cut them out altogether.

Bacteria have been performing that clever evolutionary stunt for millennia, but it wasn't until the early 2000s that food scientists at a Danish yogurt company realized just how clever the bacterial system was when they noticed that their cultures were turning too sour. They discovered that the cultures were CRISPRing invaders, altering the taste considerably. It made for bad dairy, but the scientific discovery was immediately recognized as a big one.

About a decade later, in 2012, Doudna and Charpentier tweaked the system to make it more standardized and user-friendly, and showed that not just bacterial DNA but any piece of DNA has this ability. That was a game changer. Scientists have been mucking with plant, animal and human DNA since its structure was first discovered by James Watson and Francis Crick in 1953. But altering genes, especially in deliberate, directed ways, has never been easy. "The idea of gene correction is not new at all," says June. "But before CRISPR it just never worked well enough so that people could do it routinely."

Within months of Doudna's and Charpentier's discovery, Zhang showed that the technique worked to cut human DNA at specified places. With that, genetics changed overnight. Now scientists had a tool allowing them, at least in theory, to wield unprecedented control over any genome, making it possible to delete bits of DNA, add snippets of genetic material and even insert entirely new pieces of code.

Now, that theoretical potential took shape in a remarkable array of real-world applications. CRISPR produced the first mushroom that doesn't brown, the first dogs with DNA-boosted cells giving them a comic-book-like musculature, and a slew of nutritionally superior crops that are already on their way to market. There are even efforts to use CRISPR'd mosquitoes to fight Zika and malaria.

On the human side, progress has been even more dramatic. In a lab, scientists have successfully snipped out HIV from infected human cells and demonstrated that the process works in infected mice and rats as well. They're making headway in correcting the genetic defect behind sickle-cell anemia, which stands to actually cure the disease. They're making equally promising progress in treating rare forms of genetic blindness and muscular dystrophy. And in perhaps the most controversial application of CRISPR to date, in 2016 the U.K. approved the first use of the technology in healthy human embryos for research.

At the Francis Crick Institute in London, developmental biologist Kathy Niakan is using CRISPR to try to understand one of the more enduring mysteries of human development: what goes wrong at the earliest stages, causing an embryo to die and a pregnancy to fail. To be clear, Niakan will not attempt to implant the embryos in a human; her research is experimental, and the embryos are destroyed seven days after the studies begin.

Like Niakan, June is looking for answers to one of human biology's more vexing problems: why the immune system, designed to fight disease, is nearly useless against cancer. It's an issue that's kept him up at night since 2001, when his wife, not responding to the many treatments she tried, died of ovarian cancer.

"This trial is about two things: safety and feasibility," he says. It's about testing whether it's even possible to successfully edit these immune cells to make them do—in human bodies, not a petri dish—what he wants them to do. Either way, the study will yield critical information, paving the way for eventual new treatment options that are more targeted, less brutal and far smarter against tumors than systemwide chemotherapy will ever be.

As much as has been done in 2016, this is only the beginning of a kind of medicine that stands to effectively change the course of human history. "CRISPR is an empowering technology with broad applications in both basic science and clinical medicine," says

"This is only the beginning of a kind of medicine that stands to effectively change the course of human history."

Dr. Francis Collins, director of the NIH. "It will allow us to tackle problems that for a long time we probably felt were out of our reach."

The Hurdles Ahead

Because it's so easy to use, Zhang, along with the other CRISPR pioneers, says careful thought should be given to where and how it gets employed. "For the most part I don't think we are getting ahead of ourselves with the CRISPR applications," he says. "What we need to do is really engage the public, to make sure people understand what are the really exciting potential applications and what are the immediate limitations of the technology, so we really are applying it and supporting it in the right way."

Regulatory scrutiny is a given with CRISPR, and any new tool for rewriting human DNA requires federal approval. For the current Penn trial, June got the green light from the NIH Recombinant DNA Advisory Committee, established in the 1980s to assess the safety of any first-in-humans gene-therapy trials. While there are still dangers involved in any kind of gene therapy—the changes may happen in unexpected places, for example, or the edits may have unanticipated side effects—scientists have learned more about the best way to make the genetic changes, and how to deliver them more safely. So far, animal studies show CRISPR provides enough control that unexpected negative effects are rare—at least so far.

The role of regulatory oversight is less clear when the technique is used to alter food crops. Even before June's patients get infused with CRISPR'd T cells, farmers in Argentina and Minnesota will plant the world's first gene-edited crops for market. CRISPR provides an unparalleled ability to insert almost any trait into plants—drought or pest resistance, more of this vitamin or less of that nutritional villain du jour. Dupont, for instance, is putting the finishing touches on its first drought-resistant corn, and biotech company Calyxt has created a potato that doesn't produce cancerous compounds when fried; it's also planting its first crop of soy plants modified to produce higher amounts of healthy oleic-acid fats.

These edits involve deleting or amping up existing genes—not adding new ones from other species—and the U.S. Department of Agriculture has said this kind of gene-edited food crop is not significantly different from unaltered crops and therefore does not need to be regulated differently.

In the coming months, the National Academy of Sciences is expected to issue guidelines that might address some of the challenges posed by CRISPR, focusing on how and when to proceed with developing new

25

disease treatments. The report is expected to launch much-needed discussion in the scientific community and among the public as well. Whether more regulation will eventually be required likely depends on how far scientists push the limits of their editing—and how comfortable consumers and advocacy groups are with those studies.

As CRISPR goes mainstream in medicine and agriculture, profound moral and ethical questions will arise. Few would argue against using CRISPR to treat terminal cancer patients, but what about treating chronic diseases? Or disabilities? If sickle-cell anemia can be corrected with CRISPR, should obesity, which drives so many life-threatening illnesses? Who decides where that line ought to be drawn?

Questions like these weigh heavily on June and all of CRISPR's pioneering scientists. "Having this technology enables humans to alter human evolution," says Doudna. "Thinking about all the different ways it can be employed, both for good and potentially not for very good, I felt it would be irresponsible as someone involved in the earliest stages of the technology not to get out and talk about it."

Last year, Doudna invited other leaders in genetics to a summit to address the immediate concerns about applying CRISPR to human genes. The group agreed to a voluntary temporary moratorium on using CRISPR to edit the genes of human embryos that would be inserted into a woman and brought to term, since the full array of CRISPR's consequences isn't known yet. (Any current research using human embryos, including Niakan's, is lab-only.)

For researchers like June and Niakan, Doudna and Zhang, and others, proceeding carefully with CRISPR is the only way forward. But proceed they will. The sooner more answers emerge, the sooner CRISPR can mature and begin to deliver on its promise. "There are thousands of applications for CRISPR," says June. "The sky is the limit. But we have to be careful."

Understanding the Text

1. Like the nanotechnology presented in "Rebuilding Ourselves" (p. 90), CRISPR is not being developed within a vacuum. This article identifies several scientists, the institutions they work for, and other agencies that are working to regulate and oversee this technology. Does this oversight lend credibility to CRISPR? Why or why not?
2. This selection is dense with scientific language. Identify three terms or concepts you were unfamiliar with, and explain how the author did or didn't help you to understand each one.

Reflection and Response

3. According to the article, scientists have already "snipped out HIV from infected human cells" (par. 16) and are well on their way to curing sickle-cell anemia. Think about a disease, disorder, or congenital defect you or someone you know suffers from. How might your or their lives be changed by CRISPR if a "cure" could be genetically engineered?
4. Although Dr. Francis Collins, director of the National Institutes of Health, argues that "[CRISPR] will allow us to tackle problems that for a long time we probably felt were out of our reach" (par. 20), Park cautions that it is also important to regulate this science early and clearly because, as she writes, it's so easy to use. How might this be regulated and by whom?
5. With the introduction of CRISPR, the potential for genetically altering one's baby will someday be a distinct possibility. Consider the options parents-to-be currently have when their unborn child is diagnosed with a genetic condition like a heart defect or Down syndrome. How might their options change with the advent of CRISPR? Should they change? Under what circumstances? How do you personally feel about editing the genes of an unborn child and does the purpose for that intervention matter?

Making Connections

6. At the Second International Summit on Human Genome Editing (Hong Kong, November 2018), Chinese scientist He Jiankui announced that he had used CRISPR to edit the genes of twin girls in order to make them immune to HIV, a disease their father carried. His announcement was met with international outrage from the scientific community. Research He's story and identify two or three reasons why scientists found his experiment unethical; write an essay explaining how his experiment makes manifest some of the ethical concerns expressed in "The CRISPR Pioneers."
7. Since the dawn of CRISPR, members of the scientific community have cautioned that it has the potential to usher in a new age of eugenics. Investigate the American eugenics movement of the early 20th century. What motivated those eugenicists and their supporters? Do you see connections between them and CRISPR scientists and their supporters now? What can we, or should we, learn from the way that movement unfolded?
8. The 1997 film *Gattaca* imagines a future where humans have learned how to edit and control DNA and the human genome, resulting in a society where there are inferior and superior human beings. As Alice Park tells us in this article, with the development of CRISPR, "[n]ow scientists had a tool allowing them, at least in theory, to wield unprecedented control over any genome" (par. 14). Park's article looks at the potential benefits of CRISPR, whereas the film examines the dark side of this same (although fictional) technology. Compare the motivations and results of the film to the real-world motivations and goals of CRISPR.

Rebuilding Ourselves: Ushering in an Age of Synthetic Organs and Targeted Medicine

Rohit Karnik and Robert S. Langer

Rohit Karnik, professor of mechanical engineering at MIT, has published over thirty scientific articles, many with a focus on his primary research interest of nanotechnology.° Robert Langer is a professor of chemical engineering at MIT who has published over 1,400 articles in his nearly fifty-year career; he has been awarded both the National Medal of Science and the National Medal of Technology and Innovation.

In "Rebuilding Ourselves," Karnik and Langer argue that nanotechnology has the potential to drastically change how medicine is tested and administered, giving us a future where drugs are virtually free of side effects. And it's not just the field of pharmaceuticals that will advance; nanotechnology may one day allow scientists to grow transplantable organs in the lab, eliminating long waits for an organ match and saving countless lives. Nanotechnology is constantly evolving; its possible applications as described in this article are infinite, but does the potential for groundbreaking benefits demand that we push forward with this technology?

An American male born in 1850 could expect to live to be 38. An American female had a life expectancy of 41. Today, someone who died that young would seem to have had been cut down in middle age, and it would be true: The life expectancy for American men and women has doubled. People in developing and least developed nations also have added to their expected lifespans—extending them by nearly 25 years since 1950.

One reason for the increase came from advances in medical science and technology. In 1850, for instance, scientists were still debating whether microorganisms caused disease. Once accepted, germ theory gave researchers and physicians a new way to combat illnesses. Practical applications included sophisticated vaccines and antibiotics, and such simple measures as washing hands and instruments before surgery and rigorously segregating drinking and waste water.

Today, we are in the midst of another explosion in medical technology. From powerful antivirals and chemotherapeutics to robotic surgery and genomic medicine, we are developing new ways to fight disease, carry out surgical procedures, and even transplant organs. Yet despite these advances—and sometimes because of them—we face new challenges,

Nanotechnology: technology taking place on an imperceptibly tiny (nano) scale.

ranging from antibiotic-resistant bacteria to long waiting lists for scarce donor organs.

We are slowly disentangling the complex processes of life—such as why some cells form a liver and others a heart, and how cells transport some molecules (but not others) into their interior—and starting to learn practical ways to apply this knowledge. Nearly all our bodies' biochemical processes occur at the molecular level. Nanotechnology gives us a set of tools to influence and even control these interactions. For example, nanotechnology is already being used to create systems that seek out and attack the weaknesses of cancer and infections. In the lab today, we are developing nano-enabled systems that target only tumors, so we can potentially deliver medicine only where needed.

5 Nanotechnology also has the potential to provide many of the tools needed to create artificial organs. Over the past 15 years, researchers have made enormous headway in growing pumping heart muscles and sections of functional kidneys, livers, and cartilage. Instead of waiting for transplants, physicians in the future may generate entire organs from patients' own healthy cells. The ability to regenerate body tissues and organs using a patient's own cells would also eliminate the need for a lifetime of medication to keep the body from rejecting the transplant.

> "Instead of waiting for transplants, physicians in the future may generate entire organs from patients' own healthy cells."

Scaffolds, Tissues, and Organs

Building tissues and organs is an enormously complex task. For decades, researchers tried and failed to grow even the simplest functioning organs in Petri dishes. When injected in animals, cells nearly always failed to form tissue.

This problem has begun to yield to an onslaught of basic research. To greatly simplify, one approach is to start with cells, including stem cells, which have not yet begun to develop into tissue, assemble them close together in three-dimensional structures, and prompt them with the right physical and chemical cues to develop properly. Researchers have made enormous progress over the past decade. They have, for example, produced neurons with axons and dendrites, elongated heart tissue that contracts, and functional liver tissue—all outside the body.

While researchers have generated thin sections of various tissues, replicating complex functional 3-D organs with specialized features and blood vessels to supply nutrients is still beyond our reach.

We face a number of basic challenges. We need to understand fully how cells differentiate into specialized cells and then organize themselves into the complex structures needed to carry out organ functions. We need to understand how mechanical and chemical stimuli in the environment around the cell influence the complex processes occurring within it. And we need to build on this knowledge to engineer 3-D structures that will enable cells to form tissue outside the body.

10 These 3-D structures, or scaffolds, combine multiple length scales with complex chemical signaling agents. They consist of intricately woven protein fibers (10 to 500 nanometers in diameter) that form a graded array of pores. Their surfaces are dotted with adhesive proteins to bind cells to them, and growth factors to prompt cell growth, shape, migration, and differentiation.

Reproducing natural scaffolds is not a simple task. One approach involves electrospinning. It uses an electric field to form a porous yet interconnected matrix of polymeric fibers. Researchers then add a combination of growth factors, enzymes, drugs, DNA, and RNA to signal the cells that populate the matrix.

Another approach is to create scaffolds through molecular self-assembly. This involves synthesizing molecules with different polarities or solubilities on opposite ends. Under the proper conditions, the molecules will form ordered structures spontaneously, the way a group of rod magnets will try to align north-to-south when placed on a table.

Using self-assembly, researchers have formed peptide-based 3-D gels and scaffolds with fiber diameters as small as 10 nanometers and pore sizes ranging from 5 to 200 nanometers, far more graded than those produced by electrospinning. Cells can also be embedded in these gels during the fabrication process, which is bringing us closer to realizing complex synthetic tissue structures.

Scaffolds provide the 3-D structure needed to keep cells close to one another. They also hold the adhesive molecules needed to bind the cells to the scaffold and can provide cues that guide cell development. For example, adding neural growth factors to a self-assembled scaffold causes neural progenitor cells to differentiate rapidly into neurons rather than other types of cells.

15 The topography of the scaffold also influences cell development. On smooth surfaces, for example, epithelial cells grow rounded. On surfaces with nanoscale grooves and ridges, they stretch and elongate along the grooves. Many researchers believe the patterns generate anisotropic stresses in cells that cause elongation.

Scaffolds are the building blocks of tissue, but each type of tissue requires a scaffold engineered to its own individual requirements. The heart, for example, needs a dense, elastic scaffold that forces cells to elongate and couple mechanically with one another. Without it, they will not produce aligned contractions.

Bone, on the other hand, needs a composite scaffold that includes nanocrystals of hydroxyapatite, a calcium mineral, in a matrix of proteins. Shape counts. Scaffolds with needle-shaped hydroxyapatite crystals promote bone cell differentiation better those with cylindrical or spherical crystals.

In addition to scaffold topography and chemistry, tissues present other challenges. For example, liver, pancreas, and kidney tissues secrete, absorb, and transport biochemicals. This requires a complex structure where cells face blood vessels on one side and ducts that deliver those chemicals on the other. While the polarized cells will self-organize to form ducts, it is difficult to get them to mimic the function of natural organs.

Nanotechnology can also help us alter natural designs. For example, in the laboratory, researchers have shown that it may be possible to compensate for the limitations of human-made tissues by adding carbon nanotubes. They act like a reinforcement to give synthetic tissue the strength, stiffness, and viscoelastic performance of natural membranes. Carbon nanotubes have been shown to improve the electrical responsiveness of artificially grown neurons. We can potentially use electrically or optically active nanostructures to trigger certain processes or measure how tissues develop.

We are yet a long way from creating viable replacement tissues and [20] organs. Still, we can reap significant rewards even from imperfectly grown organs by using them to test new medicines, because they go a long way beyond Petri-dish grown cells in mimicking the body's environment.

Many medicines look promising in Petri dishes but prove ineffective or too toxic in animals and animals. This is because we test them on individual cells, rather than on the complex living system of cells that make up tissues and organs in our bodies.

In 2012, the National Institutes of Health and the Defense Advanced Research Projects Agency launched a $132 million program to create a human-on-a-chip, collections of tissues that mimic interdependent organ behavior reasonably well and react more naturally when testing drugs for efficacy and toxicity. This could improve initial drug screening, reduce the cost of prolonged animal testing, and speed the development of new medications.

Nano Design of Medicine

If tissue engineering represents the promise of the future, then nanomedicine is the emerging reality of the present. Already, more than 50 pharmaceutical companies in the United States are developing nano-enabled medicines to treat, image, or diagnose cancer, and the number is growing rapidly. Nanotherapeutics to treat pain and infectious diseases are under development as well.

Nanotechnology is well-suited for delivering medications. As the famous physician, Paracelsus, noted 500 years ago, all medicines are poisons in high enough doses. Cancer medications, for example, may destroy tumors, but they kill healthy cells as well. This is why chemotherapy often produces such devastating side effects. An ideal drug would target only the diseased organ or tissue in steady, sustained doses.

Nanomaterials promise a combination of approaches that may overcome some of these limitations on drug delivery. Researchers began applying nanotechnology to these problems in a simple way more than 25 years ago. The first clinically used nanoscale systems used liposomes, small spheres made from natural fatty acids, to encapsulate cancer medicines. Doxil, a liposome carrying the chemotherapeutic doxorubicin, became the first nanocarrier approved by U.S. Food and Drug Administration in 1995 for treatment of Kaposi's sarcoma.

Liposomes measure from tens of nanometers to roughly 100 nanometers in diameter, large enough to avoid clearance by the kidneys but small enough to avoid drawing too much attention from the body's immune system. They tend to remain in the blood stream until they slip through the leaky blood vessels that usually surround tumors. Since tumors have no way to remove liposomes effectively, they stay there until they release medication that the tumor cells will absorb.

Another notable nanomedicine is Abraxane by Abraxis BioScience (which has been acquired by Celgene) and approved by the U.S. Food and Drug Administration in 2005 to treat metastatic breast cancer. It consists of an anticancer drug, paclitaxel, which could not be used without a solvent that produces severe reactions in some patients. Abraxane eliminates the solvent by coating paclitaxel with albumin, a common protein that is soluble in water. The carrier enables paclitaxel to flow through the bloodstream and accumulate in the tumor.

Doxil and Abraxane reduce side effects by encapsulating toxic drugs until they reach their destination, where they accumulate passively. Next-generation nanosystems take this one step further, peppering drug carrier surfaces with targeting ligands, chemical groups that bind to receptors that are found most frequently on tumor surfaces. Recently, BIND Biosciences performed Phase I clinical trials of the first such targeted nanoparticle drug carrier to treat prostate cancer that uses this approach to deliver the drug preferentially to tumors.

These rationally designed nanomedicines do a better job of encapsulating drugs, ranging from small molecules that would never otherwise pass through the kidneys to much larger peptides, proteins, DNA, and RNA.

They can also encapsulate more than one medicine at a time, making treatments more synergistic.

Rationally designed nanocarriers can take advantage of size and shape. For example, spherical particles 100 to 200 nanometers in diameter tend to remain in circulation longest because they are not so easily filtered by the liver or spleen. And certain elongated nanocarriers do a good job navigating the body's immune system, slipping through narrow filters and pulling away from immune system cells the way a log catches a current as it floats down a rocky stream. They may also circulate in the bloodstream longer than spherical nanoparticles.

Moreover, certain types of cells have a preference for certain shapes, and ingest them more readily than others. These nanocarriers can also be coated with "stealth" layers, so white blood cells do not recognize and attack them. This also enables medications to remain in the bloodstream longer.

Nanotechnology also enables new types of therapies that do not use drugs. Instead, we can engineer targeted nanoparticles to absorb light or electromagnetic radiation, and locally heat tissue to kill it. Such particles could concentrate in diseased tissues, and could be activated without harming healthy tissues. This would enable physicians to treat infections and tumors that resist medication and are difficult to remove surgically.

Nanoparticles could also be used as diagnostics. For example, T2 Biosystems is commercializing a magnetic nanoparticle-based platform to rapidly diagnose bacterial and fungal infections without sample purification.

Reaping the Rewards

Nanomedicine is rapidly moving into the mainstream and is poised to increasingly influence treatment of diseases such as cancer. The work of researchers has moved from laboratories to new startups and established pharmaceutical companies. And because concepts such as nanocarriers provide a flexible platform technology, they should prove easy to adapt to new breakthroughs as we unravel the biochemical secrets of our bodies.

In terms of tissue and organ engineering, we have come a long way from the days when we had little control over the growth and development of cells in Petri dishes. Today, outside the body, researchers can create heart muscle that contracts, fully differentiated nerve cells, and liver tissues with complex structures.

To be sure, we have a long way to go before we can grow fully functional 3-D organs with blood vessels, ducts, and other specialized structures running through their interior, or before we can send pharmaceuticals to kill individual disease-causing cells. But we expect to reap great rewards—and soon—as we apply nanotechnology to medicine. Nanocarriers have not yet achieved their full potential, but they are already prolonging lives and fighting infection and cancer. We may not need fully functional organ tissues to speed drug testing with organs on a chip.

Nature had billions of years to learn how to develop organs and the body's biochemical systems. But the tools of nanotechnology have given us the ability to begin to understand these designs, and the potential to adapt them to improving human life. And perhaps, one day in the not too distant future, no one will die waiting for an organ transplant.

Understanding the Text

1. Karnik and Langer claim that "[n]anotechnology is well-suited for delivering medications" (par. 24). How so? What examples do they provide to support this claim?
2. Researchers have yet to fully develop "fully functional 3-D organs" (par. 8). Why? What barriers do they face in their attempts to generate human organs?
3. Contrary to the image of Dr. Frankenstein sitting alone in his lab, this research is not a solitary business. What agencies, schools, and other entities are mentioned in the article? What does this tell you about the breadth of this research?

Reflection and Response

4. Throughout the article the authors present nanotechnology in a positive light. What is their overall argument? What is some of the language they use to indicate they are in favor of this technology? What effect does this word choice have on the reader?
5. One of the benefits of nanomedicine could be the reduction of severe side effects brought on by certain prescription drugs. Think of a negative side effect you or someone you know experienced as a result of a medication. Was the side effect worth it to treat the original problem? How might very severe side effects cause someone to forgo treatment?

Making Connections

6. The authors argue that "[n]anotechnology also enables new types of therapies that do not use drugs" (par. 32); they go on to explain that "this would enable physicians to treat infections . . . that resist medication and are difficult to remove surgically" (par. 32). Do some research on superbugs. What are they, who do they affect, and how are these infections currently treated? Now that you have read about nanomedicine, do you think it has the potential to treat superbugs? Explain your answer using evidence from your outside research.

7. In the future, scientists may be able to create artificial organs via nanotechnology. As Karnik and Langer point out, this developing technology could be life-saving for people on the wait-list for an organ transplant. Do some research on the current state of organ donation: what organs are in highest demand, how long is the average wait, and what are some alternatives, both legal and illegal, to the wait-list? Compare and contrast the risks and benefits that come with each alternative.
8. Companies, pharmaceutical and otherwise, have a long history of testing their products on animals. Nanotechnology has the potential to change this, by providing tissues that "mimic interdependent organ behavior reasonably well and react more naturally when testing drugs for efficacy and toxicity" (par. 22). Research some of the ethical and accuracy challenges associated with testing pharmaceuticals on animals. How might nanotechnology, like human organs-on-a-chip or artificial organs, address some of those concerns?

The Healthcare Challenge; from *Rise of the Robots: Technology and the Threat of a Jobless Future*

Martin Ford

Martin Ford is a computer engineer, entrepreneur, and futurist with a specific interest in artificial intelligence, or AI. In his work, he often investigates the impacts of automation and AI on the economy. He has authored three books on these topics and written for numerous publications including the *New York Times*, *The Atlantic*, and *Forbes*.

In this excerpt from his best-selling book *Rise of the Robots* (2015), Ford looks at the various challenges doctors face in a world where technology and communication are constantly developing, and the solutions that AI may provide. Despite the fact that they have more technological resources than ever before, doctors often still operate within information silos. Ford examines the potential benefits of using IBM's Watson technology, a form of AI, in diagnosing and treating patients. Although an A.I.'s diagnosis may at first sound impersonal or even frightening, what might we gain from a technology that can access, evaluate, and act on more information than any doctor could ever possibly understand on his or her own?

In May 2012, a fifty-five-year-old man checked into a clinic at the University of Marburg in Germany. The patient suffered from fever, an inflamed esophagus, low thyroid hormone levels, and failing vision. He had visited a series of doctors, all of whom were baffled by his condition. By the time he arrived at the Marburg clinic, he was nearly blind and was on the verge of heart failure. Months earlier, and a continent away, a very similar medical mystery had culminated with a fifty-nine-year-old woman receiving a heart transplant at the University of Colorado Medical Center in Denver.

The answer to both mysteries turned out to be the same: cobalt poisoning.[1] Both patients had previously received artificial hips made from metal. The metal implants had abraded over time, releasing cobalt particles and exposing the patients to chronic toxicity. In a remarkable coincidence, papers describing the two cases were published independently in two leading medical journals on nearly the same day in February 2014. The report published by the German doctors came with a fascinating twist: whereas the American team had resorted to surgery, the German team had managed to solve the mystery not because of their training but because one of the doctors had seen a February 2011 episode of the television show *House*. In the episode, the show's protagonist, Dr. Gregory

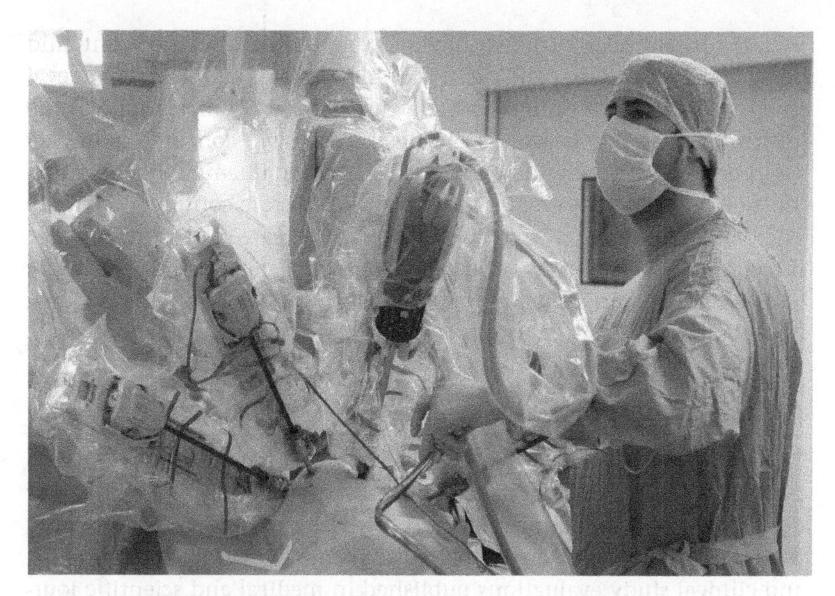

The Da Vinci surgical system allows surgeons to perform minimally invasive and complex surgeries through the use of the da Vinci robot's "hands," which are far smaller than that of a human and more accurate due to the enhanced visual capabilities of the system.
Media for Medical/UIG/Getty Images

House, is faced with the same problem and makes an ingenious diagnosis: cobalt poisoning resulting from a metal prosthetic hip replacement.

The fact that two teams of doctors can struggle to make the same diagnosis—and that they can do so even when the answer to the mystery has been broadcast to millions of prime-time television viewers—is a testament to the extent to which medical knowledge and diagnostic skill are compartmentalized in the brains of individual physicians, even in an age when the internet has enabled an unprecedented degree of collaboration and access to information. As a result, the fundamental process that doctors use to diagnose and treat illnesses has remained, in important ways, relatively unchanged. Upending that traditional approach to problem solving, and unleashing all the information trapped in individual minds or published in obscure medical journals, likely represents one of the most important potential benefits of artificial intelligence and big data as applied to medicine.

In general, the advances in information technology that are disrupting other areas of the economy have so far made relatively few inroads into the health care sector. Especially hard to find is any evidence that technology is resulting in meaningful improvements in overall efficiency. In 1960, health care represented less than 6 percent of the US economy.[2]

By 2013 it had nearly tripled, having grown to nearly 18 percent, and per capita health care spending in the United States had soared to a level roughly double that of most other industrialized countries. One of the greatest risks going forward is that technology will continue to impact asymmetrically, driving down wages or creating unemployment across most of the economy, even as the cost of health care continues to climb. The danger, in a sense, is not too many health care robots but too few. If technology fails to rise to the health care challenge, the result is likely to be a soaring, and ultimately unsustainable, burden on both individual households and the economy as a whole.

Artificial Intelligence in Medicine

The total amount of information that could potentially be useful to a physician attempting to diagnose a particular patient's condition or design an optimal treatment strategy is staggering. Physicians are faced with a continuous torrent of new discoveries, innovative treatments, and clinical study evaluations published in medical and scientific journals throughout the world. For example, MEDLINE, an online database maintained by the U.S. National Library of Medicine, indexes over 5,600 separate journals—each of which might publish anywhere from dozens to hundreds of distinct research papers every year. In addition, there are millions of medical records, patient histories, and case studies that might offer important insights. According to one estimate, the total volume of all this data doubles roughly every five years.[3] It would be impossible for any human being to assimilate more than a tiny fraction of the relevant information even within highly specific areas of medical practice.

As we saw in Chapter 4, medicine is one of the primary areas where IBM foresees its Watson technology having a transformative impact. IBM's system is capable of churning through vast troves of information in disparate formats and then almost instantly constructing inferences that might elude even the most attentive human researcher. It's easy to imagine a near-term future where such a diagnostic tool is considered indispensable, at least for physicians confronting especially challenging cases.

The MD Anderson Cancer Center at the University of Texas handles over 100,000 patients at its Houston hospital each year and is generally regarded as the best cancer treatment facility in the United States. In 2011, IBM's Watson team began working with MD Anderson's doctors to build a customized version of the system geared toward assisting oncologists working with leukemia cases. The goal is to create an interactive adviser capable of recommending the best evidence-based treatment options, matching patients with clinical drug trials, and highlighting possible dangers or side effects that might threaten specific patients. Initial progress on the project proved to be somewhat slower than the team expected, largely because of the challenges associated with designing algorithms capable of taking on the complexities of cancer diagnosis and treatment. Cancer, it turns out, is tougher than *Jeopardy!* Nonetheless, by January 2014, the *Wall Street Journal* reported that the Watson-based leukemia system at MD Anderson was "back on track" toward becoming operational.[4] Researchers hope to expand the system to handle other kinds of cancer within roughly two years. It's very likely that the lessons IBM takes away from this pilot program will enable the company to streamline future implementations of the Watson technology.

Once the system is operating smoothly, the MD Anderson staff plans to make it available via the internet so that it can become a powerful resource for doctors everywhere. According to Dr. Courtney DiNardo, a leukemia expert, the Watson technology has the "potential to democratize cancer care" by allowing any physician to "access the latest scientific knowledge and MD Anderson's expertise." "For physicians who aren't leukemia experts," she added, the system "can function as an expert second opinion, allowing them to access the same knowledge and information" relied on by the nation's top cancer treatment center. DiNardo also believes that, beyond offering advice for specific patients, the system "will provide an unparalleled research platform that can be used to generate questions, explore hypotheses and provide answers to critical research questions."[5]

> "The Watson technology has the 'potential to democratize cancer care' by allowing any physician to 'access the latest scientific knowledge.'"

Watson is currently the most ambitious and prominent application of artificial intelligence to medicine, but there are other important success stories as well. In 2009, researchers at the Mayo Clinic in Rochester, Minnesota, built an artificial neural network designed to diagnose cases of endocarditis—an inflammation of the inner layer of the heart. Endocarditis normally requires that a probe be inserted into the patient's esophagus in order to determine whether or not the inflammation is caused by a potentially deadly infection—a procedure that is uncomfortable, expensive, and itself carries risks for the patient. The Mayo doctors instead trained a neural network to make the diagnosis based on routine tests and observable symptoms alone, without the need for the invasive technique. A study involving 189 patients found that the system was accurate more than 99 percent of the time and successfully saved over half of the patients from having to needlessly undergo the invasive diagnostic procedure.[6]

10

One of the most important benefits of artificial intelligence in medicine is likely to be the avoidance of potentially fatal errors in both diagnosis and treatment. In November 1994, Betsy Lehman, a thirty-nine-year-old mother of two and a widely read columnist who wrote about health-related issues for the *Boston Globe*, was scheduled to begin her third round of chemotherapy as she continued her battle against breast cancer. Lehman was admitted to the Dana-Farber Cancer Institute in Boston, which, like MD Anderson, is regarded as one of the country's preeminent cancer centers. The treatment plan called for Lehman to be given a powerful dose of cyclophosphamide—a highly toxic drug intended to wipe out her cancer cells. The research fellow who wrote the medication order made a simple numerical error, which meant that the total dosage Lehman received was about four times what the treatment plan actually called for. Lehman died from the overdose on December 3, 1994.[7]

Lehman was just one of as many as 98,000 patients who die in the United States each year as a direct result of preventable medical errors.[8] A 2006 report by the U.S. Institute of Medicine estimated that at least 1.5 million Americans are harmed by medication errors alone, and that such mistakes result in more than $3.5 billion in additional annual treatment costs.[9] An AI system with access to detailed patient histories, as well as information about medications, including their associated toxicity and side effects, would potentially be able to prevent errors even in very complex situations involving the interaction of multiple drugs. Such a system could act as an interactive adviser to doctors and nurses, offering instantaneous verification of both safety and effectiveness before medication is administered, and—especially in situations where hospital staff are tired or distracted—it would be very likely to save both lives and needless discomfort and expense.

Once medical applications of artificial intelligence evolve to the point where the systems can act as true advisers capable of providing consistently high-quality second opinions, the technology could also help rein in the high costs associated with malpractice liability. Many physicians feel the need to practice "defensive medicine" and order every conceivable test in an attempt to protect themselves against potential lawsuits. A documented second opinion from an AI system versed in best practice standards could offer doctors a "safe harbor" defense against such claims. The result might be less spending on needless medical tests and scans as well as lower malpractice insurance premiums.

Looking even further ahead, we can easily imagine artificial intelligence having a genuinely transformative impact on the way medical services are delivered. Once machines demonstrate that they can offer accurate diagnosis and effective treatment, perhaps it will not be necessary for a physician to directly oversee every encounter with every patient.

In an op-ed I wrote for the *Washington Post*, shortly after Watson's 2011 triumph at playing *Jeopardy!*, I suggested that there may eventually be an opportunity to create a new class of medical professionals: persons educated with perhaps a four-year college or master's degree, and who are trained primarily to interact with and examine patients—and then to convey that information into a standardized diagnostic and treatment system.[10] These new, lower-cost practitioners would be able to take on many routine cases, and could be deployed to help manage the dramatically growing number of patients with chronic conditions such as obesity and diabetes.

Physicians groups would, of course, be likely to oppose the influx of these less-educated competitors. However, the reality is that the vast majority of medical school graduates are not especially interested in entering family practice, and they are even less excited about serving rural areas of the country. Various studies predict a shortage of up to 200,000 doctors within the next fifteen years as older doctors retire, the Affordable Care Act plan brings as many as 32 million new patients into the health insurance system, and an aging population requires more care.[11] The shortage will be most acute among primary-care physicians as medical school graduates, typically burdened by onerous levels of student debt, choose overwhelmingly to enter more lucrative specialties.

These new practitioners, trained to utilize a standardized AI system that encapsulates much of the knowledge that doctors acquire during the course of nearly a decade of intensive training, could handle routine cases, while referring patients who require more specialized care to physicians. College graduates would benefit significantly from the availability of a compelling new career path, especially as intelligent software increasingly erodes opportunities in other sectors of the job market.

In some areas of medicine, particularly those that don't require direct interaction with patients, advances in AI are poised to drive dramatic productivity increases and perhaps eventually full automation. Radiologists, for example, are trained to interpret the images that result from various medical scans. Image processing and recognition technology is advancing rapidly and may soon be able to usurp the radiologist's traditional role. Software can already recognize people in photos posted on Facebook and even help identify potential terrorists in airports. In September 2012, the FDA approved an automated ultrasound system for screening women for breast cancer. The device, designed by U-Systems, Inc., is designed to help identify cancer in the roughly 40 percent of women whose dense breast tissue can render standard mammogram technology ineffective. Radiologists still need to interpret the images, but doing so now takes only about three minutes. That compares with twenty to thirty minutes for images produced using standard handheld ultrasound technology.[12]

Automated systems can also provide a viable second opinion. A very effective—but expensive—way to increase cancer detection rates is to have two radiologists read every mammogram image separately and then reach a consensus on any potential anomalies identified by either doctor. This "double reading" strategy results in significantly improved cancer detection and also dramatically reduces the number of patients who have to be recalled for further testing. A 2008 study published in the *New England Journal of Medicine* found that a machine can step into the role of the second doctor. When a radiologist is paired with a computer-aided detection system, the results are just as good as having two doctors separately interpret the images.[13]

Pathology is another area where artificial intelligence is already encroaching. Each year, over a hundred million women throughout the world receive a Pap test to screen for cervical cancer. The test requires that cervical cells be deposited on a glass microscope slide and then be examined by a technician or doctor for signs of malignancy. It's a labor-intensive process that can cost up to $100 per test. Many diagnostic labs, however, are now turning to a powerful automated imaging system manufactured by BD, a New Jersey–based medical device company. In a 2011 series of articles about job automation for *Slate,* technology columnist Farhad Manjoo called the BD FocalPoint GS Imaging System "a marvel of medical engineering" whose "image-searching software rapidly scans slides in search of more than 100 visual signs of abnormal cells." The system then "ranks the slides according to the likelihood they contain disease" and finally "identifies 10 areas on each slide for a human to scrutinize."[14] The machine does a significantly better job of finding instances of cancer than human analysts alone, even as it roughly doubles the speed at which the tests can be processed.

Hospital and Pharmacy Robotics

The pharmacy at the University of California Medical Center in San Francisco prepares about 10,000 individual doses of medication every day, and yet a pharmacist never touches a pill or a medicine bottle. A massive automated system manages thousands of different drugs and handles everything from storing and retrieving bulk pharmaceutical supplies to dispensing and packaging individual tablets. A robotic arm continuously picks pills from an array of bins and places them in small plastic bags. Every dose goes into a separate bag and is labeled with a barcode that identifies both the medication and the patient who should receive it. The machine then arranges each patient's daily meds in the order that they need to be taken and binds them together. Later, the nurse who administers the medication will scan the barcodes on both the dosage bag and the patient's wrist band. If they don't match, or if the medication is being given at the wrong time, an alarm sounds. Three other specialized robots automate the preparation of injectable medicines; one of these robots deals exclusively with highly toxic chemotherapy drugs. The system virtually eliminates the possibility of human error by cutting humans almost entirely out of the loop.

UCSF's $7 million automated system is just one of the more spectacular examples of the robotic transformation that's unfolding in the pharmacy industry. Far less expensive robots, not much larger than a vending machine, are invading retail pharmacies located in drug and grocery stores. Pharmacists in the United States require extensive training (a four-year doctoral degree) and have to pass a challenging licensing exam. They are also well paid, earning about $117,000 on average in 2012. Yet, especially in retail settings, much of the work is fundamentally routine and repetitive, and the overriding concern is to avoid a potentially deadly mistake. In other words, much of what pharmacists do is almost ideally suited to automation.

Once a patient's medication is ready to leave a hospital pharmacy, it's increasingly likely that it will do so in the care of a delivery robot. Such machines already cruise the hallways in huge medical complexes delivering drugs, lab samples, patient meals, or fresh linens. The robots can navigate around obstacles and use elevators. In 2010, El Camino Hospital in Mountain View, California, leased nineteen delivery robots from Aethon, Inc., at an annual cost of about $350,000. According to one hospital administrator, paying people to do the same work would have cost over a million dollars per year.[15] In early 2013, General Electric announced plans to develop a mobile robot capable of locating, cleaning, sterilizing, and delivering the thousands of surgical tools used in operating rooms. The tools would be tagged with radio-frequency identification (RFID) locator chips, making it easy for the machine to find them.[16]

Beyond the specific areas of pharmacy and hospital logistics and delivery, autonomous robots have so far made relatively few inroads. Surgical robots are in widespread use, but they are designed to extend the capabilities of surgeons, and robotic surgery actually costs more than traditional methods. There is some preliminary work being done on building more ambitious surgical robots; for example, the I-Sur project is an EU-backed consortium of European researchers who are attempting to automate basic procedures like puncturing, cutting, and suturing.[17] Still, for the foreseeable future, it seems inconceivable that any patient would be allowed to undergo an invasive procedure without a doctor being present and ready to intervene, so even if such technology materializes, any cost savings would likely be marginal at best.

While recent applications of AI and robotics to the health care field are impressive and advancing rapidly, they are, for the most part, just beginning to nibble at the edges of the hospital cost problem. With the exception of pharmacists, and possibly doctors or technicians who specialize in analyzing images or lab specimens, automating even a significant portion of the jobs done by most skilled health care workers remains a daunting challenge. For those seeking a career that is likely to be relatively safe from automation, a skilled health care profession that requires direct interaction with patients remains an excellent bet. That calculus could, of course, change in the more distant future. Twenty or thirty years from now, I think, it's impossible to say with any real confidence what might be technologically possible.

Notes

1. These two cases of cobalt poisoning were reported by Gina Kolata, "As Seen on TV, a Medical Mystery Involving Hip Implants Is Solved," *New York Times,* February 6, 2014, http://www.nytimes.com/2014/02/07/health/house-plays-a-role-in-solving-a-medical-mystery.html.
2. Catherine Rampell, "U.S. Health Spending Breaks from the Pack," *New York Times* (Economix blog), July 8, 2009, http://economix.blogs.nytimes.com/2009/07/08/us-health-spending-breaks-from-the-pack/.
3. IBM corporate website, http://www-03.ibm.com/innovation/us/watson/watson_in_healthcare.shtml.
4. Spencer E. Ante, "IBM Struggles to Turn Watson Computer into Big Business," *Wall Street Journal,* January 7, 2014, http://online.wsj.com/news/articles/SB10001424052702304887104579306881917668654.
5. Dr. Courtney DiNardo, as quoted in Laura Nathan-Garner, "The Future of Cancer Treatment and Research: What IBM Watson Means for Our Patients," *MD Anderson—Cancerwise,* November 12, 2013, http://www2.mdanderson.org/cancerwise/2013/11/the-future-of-cancer-treatment-and-research-what-ibm-watson-means-for-patients.html.
6. Mayo Clinic Press Release: "Artificial Intelligence Helps Diagnose Cardiac Infections," September 12, 2009, http://www.eurekalert.org/pub_releases/2009–09/mc-aih090909.php.
7. National Research Council, *Preventing Medication Errors: Quality Chasm Series* (Washington, DC: National Academies Press, 2007), p. 47.
8. National Research Council, *To Err Is Human: Building a Safer Health System* (Washington, DC: National Academies Press, 2000), p. 1.

9. National Academies News Release: "Medication Errors Injure 1.5 Million People and Cost Billions of Dollars Annually," July 20, 2006, http:// www8.nationalacademies.org/onpinews/newsitem.aspx?RecordID=11623.
10. Martin Ford, "Dr. Watson: How IBM's Supercomputer Could Improve Health Care," *Washington Post,* September 16, 2011, http://www.washingtonpost.com/opinions/dr-watson-how-ibms-supercomputer-could-improve-health-care/2011/09/14/gIQAOZQzXK_story.html.
11. Roger Stark, "The Looming Doctor Shortage," Washington Policy Center, November 2011, http://www.washingtonpolicy.org/publications/notes/looming-doctor-shortage.
12. Marijke Vroomen Durning, "Automated Breast Ultrasound Far Faster Than Hand-Held," *Diagnostic Imaging,* May 3, 2012, http://www.diagnosticimaging.com/articles/automated-breast-ultrasound-far-faster-hand-held.
13. On the "double reading" strategy in radiology, see Farhad Manjoo, "Why the Highest-Paid Doctors Are the Most Vulnerable to Automation," *Slate,* September 27, 2011, http://www.slate.com/articles/technology/robot_invasion/2011/09/will_robots_steal_your_job_3.html; I. Anttinen, M. Pamilo, M. Soiva, and M. Roiha, "Double Reading of Mammography Screening Films—One Radiologist or Two?," *Clinical Radiology* 48, no. 6 (December 1993): 414–421, http://www.ncbi.nlm.nih.gov/pubmed/8293648?report=abstract; and Fiona J. Gilbert et al., "Single Reading with Computer-Aided Detection for Screening Mammography," *New England Journal of Medicine,* October 16, 2008, http://www.nejm.org/doi/pdf/10.1056/NEJMoa0803545.
14. Manjoo, "Why the Highest-Paid Doctors Are the Most Vulnerable to Automation."
15. Rachael King, "Soon, That Nearby Worker Might Be a Robot," *Bloomberg Businessweek,* June 2, 2010, http://www.businessweek.com/stories/2010-06-02/soon-that-nearby-worker-might-be-a-robotbusinessweek-business-news-stock-market-and-financial-advice.
16. GE Corporate Press Release: "GE to Develop Robotic-Enabled Intelligent System Which Could Save Patients Lives and Hospitals Millions," January 30, 2013, http://www.genewscenter.com/Press-Releases/GE-to-Develop-Robotic-enabled-Intelligent-System-Which-Could-Save-Patients-Lives-and-Hospitals-Millions-3dc2.aspx.
17. I-Sur website, http://www.isur.eu/isur/.

Understanding the Text

1. Why is it so difficult for doctors to remain up-to-date on medical discoveries and new treatments? What examples does Ford provide that illustrate this challenge?
2. This article centers on the AI developed by IBM and deployed as Watson. What is Watson? Some outside research may be required to answer this question.
3. Identify three areas where AI has the potential to benefit the health-care industry.

Reflection and Response

4. In 2011, Ford suggested in an op-ed for *The Washington Post* that there may eventually be an opportunity "to create a new category of medical professionals" who have less formal education than a traditional MD, but who gather information and assess symptoms so they can relay them to an AI "practitioner" for diagnosis. Would you, as a patient, feel comfortable with this method of care? Consider how this method is different from that of traditional doctors relying on technology like CT scans to come to a diagnosis.
5. Can you identify any area(s) where Ford implies or explicitly states an opinion about the use of AI in health care? Does this affect the overall value of the piece?

Making Connections

6. Research the use of Watson in health care since the publication of *Rise of the Robots*. Does Watson still show the promise that Ford described in this excerpt from the book? What contributions to health care has Watson's technology made? Develop an argument about its current and potential value.
7. Ford notes that the use of AI may have an economic impact on the health-care industry. Tasks once completed by humans could be performed by robots. While this might mean more efficient task completion, it might mean fewer jobs. Research the real-world impact of robots in the health-care industry. Are there significant job losses and if so, do the benefits of the technology outweigh those losses? Explain.
8. In his article "Technology Will Replace 80% of What Doctors Do" (*Fortune* magazine, 2012), author Vinod Khosla claimed that long-held medical beliefs that informed standard practices were often found to be based on incorrect assumptions and were even dangerous. This premise seems to be supported by Ford's claims that medical errors and misdiagnoses are more common than we may think. Compare and contrast three standard practices that have been reconsidered or changed as a result of technology.

Creating Good from Immoral Acts; from *Stem Cell Dialogues*

Sheldon Krimsky

Sheldon Krimsky is the Lenore Stern Professor of Humanities and Social Sciences in the Department of Urban and Environmental Planning at Tufts University. He holds a master's degree in physics and a doctorate in philosophy. He has authored fourteen books and over 200 essays and articles, with a primary focus on the connections among science, technology, ethics, and public policy. Krimsky has also served on the advisory and editorial boards of multiple scientific journals and organizations.

In the following excerpt from his book *Stem Cell Dialogues*, Krimsky uses the literary format of a dialogue, similar to one used by Plato in his Socratic dialogues, to pose several questions about how we might all define morally acceptable scientific research. In this sixth dialogue of twenty-five, his fictional characters, Franklin and Baum, debate what the word *ethical* means when it comes to the use of stem cells, whether their provenance should matter, and how one's own culture shapes how they might use them. Krimsky's piece encourages readers to question the ethical value of stem cell science and its methods on every level.

The history of Western ethics has been largely guided by two grand theories: deontology (represented by the ethics of Immanuel Kant°) and utilitarianism (represented by the ethics of John Stuart Mill°). Deontology is based on the fundamental rightness or wrongness of an action, whereas utilitarianism looks at the total balance of good versus evil when evaluating an act.

In the wake of heinous crimes committed by the Nazis during World War II, German science was critically examined and deemed responsible as an enabler, contributor, and participant. Jews, Gypsies, the mentally challenged, and other minorities living in Germany or German-occupied territories were forced to participate in unethical experiments, which have since been examined in books like *The Nazi Doctors* by Robert J. Lifton and *Racial Hygiene: Medicine Under the Nazis* by Robert Proctor.

Some members of the postwar scientific community believed that published works by German scientists who collaborated with the Nazi regime should never be used or cited in the scientific literature. Data published

Immanuel Kant: (1724–1804), German philosopher famous for developing The Categorical Imperative.
John Stuart Mill: (1806–1873), British philosopher known for his book *Utilitarianism*.

from unethical experiments, it was argued, could not be trusted, and scientists who participated in Nazi war crimes should not be honored by having their published work incorporated into the edifice of science. Besides, their work would never be cited or replicated, and thus was not reliable.

"Can one benefit from the results of what some believe to be a past immoral act without becoming complicit in the act?"

But what if some of these unethical experiments produced unique data, which no one else had, that could save lives? Is it ethical to create good out of immoral science? Stephen Post° wrote: "Because the Nazi experiments on human beings were so appallingly unethical, it follows, prima facie, that the use of their results is unethical."[1] He argues that science must draw a line between civilization and the moral abyss (the *summum malum*) around which ethics builds fences.

5 The issue of Nazi science may be an extreme example in considering whether evil acts should be exploited for any good they can produce. In the debate about stem cells and the destruction of embryos, some scientists began to raise analogies with extreme cases. If living human embryos have some ethical status, even if not fully as persons, can destroying them be justified for the common good? Can one benefit from the results of what some believe to be a past immoral act without becoming complicit in the act?[2]

• • •

Scene: Dr. Franklin flies to Germany to discuss the politics and ethics of stem cells with scientist Gordon Baum. Franklin is interested in how German scientists view their responsibility to the law and ethical norms while they investigate the medical benefits of human embryonic stem cells. She wants to understand why Germany does not permit scientists to destroy embryos in order to derive human embryonic stem cells, but allows them to work with embryonic stem cells produced this way outside the country.

FRANKLIN: Dr. Baum, there are clearly sharp legal and moral divisions over the use of human embryonic stem cells. Some countries, like Germany and the United States, have prohibitions against destroying human embryos for research, including as

Stephen Post: (b. 1951), director for the Center of Medical Humanities, Compassionate Care, and Bioethics at Stony Brook University in New York; has written about the ethical challenges of using data obtained from Nazi experiments.

source materials for embryonic stem cells. Other countries, like the United Kingdom, allow research on and destruction of embryos prior to fifteen weeks. You live in a country where it is illegal to destroy human embryos to obtain embryonic stem cells. If that is considered illegal and possibly immoral, why would it be legal or moral to work with stem cells taken from embryos destroyed in countries other than Germany?

BAUM: Germany, my country, has a reprehensible history of eugenics during the dark period of the Nazi regime. This regime selected certain people and certain fetuses as undesirable. So as a nation we are extremely sensitive about selecting or destroying embryos. Other countries are not burdened with this historical legacy and so are not as morally vigilant. Germany has chosen to create moral buffer zones to prevent the society from getting too close to embryo selection or destruction. So we can feel comfortable using hESCs derived by other countries because our experiments do not involve embryo research, selection, or destruction.

FRANKLIN: Let us assume, for the purpose of our discussion, that destroying embryos is wrong. If you as a scientist could use the products of destroyed embryos for good purposes, would you consider that ethical? Is it wrong to create good from immoral acts committed by others?[3]

BAUM: I guess there are two issues underlying your question. Can we do good things through immoral acts? And should we use the outcome of bad acts to do good things? The answer to the first question is most assuredly yes. We tell lies or violate a law (for example, pass a red light) if we know it will save someone's life. Utilitarian ethics dictates the ethical outcome of such a choice. Ethicists use many examples to illustrate how, by violating a moral norm to protect another person, we are doing the morally correct thing. The acts in question are usually specific in time and place, not continuous activities, such as telling a lie to an intruder to protect a family member. But when there is a continuous activity that violates legal or moral rules, the issue becomes more complex, and utilitarian ethics might not produce the best outcome. 10

FRANKLIN: Can you be specific?

BAUM: A continuous act would be something like torture. It is possible to justify a single act of torture on utilitarian grounds, based on the number of lives saved. But if we were to generalize the use of torture, it would clearly be immoral.

FRANKLIN: Aren't you using Kant's categorical imperative?

BAUM: Precisely! It is not possible to consistently universalize a maxim such as: "I can use torture whenever I believe the act will produce more good than evil."

FRANKLIN: Your second question was "Should we use the outcome of bad acts to do good things?" 15

BAUM: We need to look at the cases. For example, I have a physicist friend in the United States who had cancer as a young man in the 1950s. Not much was known about shielding the body from radiation therapy. German scientists under the Nazi regime did immoral experiments and produced the only data available at the time for determining the amount of lead shielding needed to protect the testes during radiation therapy. My friend used the results of these immoral experiments to advise physicians on the proper radiation shield, and he was able to have a healthy child after his treatment.

FRANKLIN: Many people believe it is wrong to buy products made by exploited child laborers. Suppose a nonprofit organization that houses, feeds, and clothes the homeless had an opportunity to purchase a crate of clothing at a very low price, but the clothes were manufactured by child labor. Would it be ethical to purchase them and distribute them to poor families?

BAUM: The critical question is whether you become an enabler of an immoral workplace. Your purchase of those products, no matter what good they do, provides support for an immoral and illegal operation. The outcome might be different if the clothes known to have been manufactured with child labor were found in an abandoned warehouse. In that case you could argue that your action to distribute the clothes would not support the illegal activities.

FRANKLIN: Now we are left with the question: If there are no universal norms or international laws preventing the destruction of embryos, only laws in distinct nation-states, is it immoral for you to use imported human embryonic stem cells that have been acquired from destroyed embryos in another country?

BAUM: The answers may differ depending on whether you are viewing the question prospectively or retrospectively. Jonas Salk's polio vaccine was tested on mentally retarded orphans without reasonable informed consent. We would not permit these experiments today. But no reasonable person would refuse the benefits of polio vaccine based on the ethical 20 transgressions that took place in the 1950s and 1960s, before we had regulations on informed consent in medical experiments.[4]

FRANKLIN: Looking prospectively, if killing embryos is immoral, then is it ethically justifiable for scientists to use human embryonic stem cells from a black market source for research designed to find the cause of and treatment for diseases?

BAUM: Let me first respond by saying that I object to the word "killing" applied to a few cells—a zygote—so early in development that they exhibit no human qualities. If there were an international consensus on embryo creation and destruction for research, then I would be the first to say that it would be immoral to accept embryonic stem cells on the black market. Without an international convention, I do not consider it immoral to use the products of fertilized human eggs a few weeks into their development, when obtained according to legal and ethical procedures in a country that allows stem cells to be sent to other countries, even countries that prohibit destroying embryos.

FRANKLIN: Let us suppose there was an international treaty that prohibited the production and destruction of embryos for research. Some have argued that using stem cells from embryos does not constitute an endorsement of their destruction. As an analogy, imagine that environmentally insensitive strip miners are engaged in practices that unnecessarily kill hundreds of fish. If nearby residents eat the fish instead of letting them rot, the residents are not endorsing the fish-killing practices. They are simply finding a positive outcome from an ecological tragedy.

BAUM: Again, we need to distinguish cases that are one-of-a-kind from those that describe continuous events. We also have to acknowledge those cases where people who acquire some benefit from an immoral act are explicitly or implicitly reinforcing the continuation of the act, where they become enablers. The high demand for embryonic stem cells, like the high demand for illicit drugs, reinforces the production of those cells and therefore is a tacit endorsement of embryo destruction. The fish story does not help us through the moral quagmire. Research on donated embryonic stem cells, harvesting embryonic stem cells, and destroying embryos in order to obtain stem cells are all part of the same enterprise. The researcher is not like a person who searches the

Dumpster and finds clothes made by underage workers. We cannot avoid the core ethical problem—whether it is always immoral to destroy a human embryo—by hiding behind the supply chain.

FRANKLIN: The IVF industry produces tens of thousands of excess embryos. Many will be destroyed when they are no longer needed. So why not use them for research? 25

BAUM: The reason for our policy in Germany is that our scientists do not want to participate in the supply chain of embryo destruction. But once the embryos are destroyed elsewhere, we will do good science with the stem cells. From a moral standpoint, for the pure researcher, the intention of those who destroyed the embryo is irrelevant.

FRANKLIN: In the United States, two sets of ethical guidelines prevail, one for federally funded research and another for privately funded research. And yet the government under the Bush administration did not allow public funding for studying human embryonic stem cells derived in the private sector. Does this seem like a consistent and viable moral position?

BAUM: The crux of the matter is whether you believe that, regardless of societal benefits, destroying an embryo created for research purposes is prima facie wrong, whether legal or not. If that is the accepted belief, then one cannot justify using human embryonic stem cells obtained in this way. Following this logic, if a scientist used human embryonic stem cells and saved a life, then the result would have to be evaluated by head-to-head comparison with other immoral acts committed to save a life.

FRANKLIN: What if an embryo is considered a person?

BAUM: Then the moral equation changes. It is universally accepted that one person cannot be sacrificed to save another. All civilized societies have embraced the principle of Immanuel Kant that a person is an "end-in-him/herself" and not a means to some other end. Personally, I cannot accept the idea that a zygote less than fifteen days old fulfills the concept of personhood. 30

FRANKLIN: There are over 100,000, some say as many as 400,000, human embryos stored under cryogenic conditions. The vast majority of these fertilized eggs, which are held for family building, will eventually not be used for IVF.[5] They will either be selectively destroyed or over time will no longer be viable. Is there any ethical reason these embryos should not be donated for research?

BAUM: If destroying an embryo is intrinsically immoral, then donating it to research will not redeem the act of its destruction. We donate our organs after death, but it would be immoral to kill a person before they die, even if their death is imminent, to harvest their organs. Because I do not believe it is intrinsically immoral to harvest stem cells from a week-old embryo, I cannot accept the conclusions that follow from that premise.

FRANKLIN: Suppose someone believes that the moral worth of embryos is not absolute, but still believes it is immoral to destroy them for harvesting stem cells. Yet they take a different stance on the use of externally fertilized embryos for reproduction. They see the moral status of the embryo differently in the context of research and reproduction. If that is a defensible moral position, then after the cryogenically preserved embryos are released from their use in IVF and await destruction, why shouldn't they be used to harvest embryonic stem cells?

BAUM: This case seems as close as you can get to living tissue donation. Even those who make a distinction between embryos for research and embryos for reproduction seem to believe that "throwaway embryos" should be available, at the consent of the donor, for contributing to human good through research. And yet there are bioethicists who would disagree with me. A Catholic priest, asked by a parishioner what she should do with her excess IVF embryos, replied that she should either put them up for adoption or pay their rent (in the freezer) for life.

FRANKLIN: But the frozen embryos will eventually die. 35

BAUM: Yes, but under Catholic teachings, allowing an embryo to die a natural death is morally distinct from destroying it to harvest its stem cells. The outcome may be the same, but the intentions are different.

FRANKLIN: If our moral respect for embryos is really a surrogate for the respect we are supposed to have for people, using doomed embryos for research seems less likely to diminish respect for human life than simply routinely discarding them in fertility clinics. Almost any way of putting embryos to use in research sends a much better message than just discarding them.[6]

Notes

1. Stephen G. Post, "The Echo of Nuremberg: Nazi Data and Ethics," *Journal of Medical Ethics* 17 (1991): 43.
2. President's Council on Bioethics, *Monitoring Stem Cell Research,* a Report of the President's Council (Washington, D.C.: U.S. Government Printing Office, January 2004), 33.
3. Howard J. Curzer, "The Ethics of Embryonic Stem Cell Research," *Journal of Medicine and Philosophy* 29 (5) (2004): 536.
4. Ibid.
5. Nicholas Wade, "Clinics Hold More Embryos Than Had Been Thought," *New York Times,* May 9, 2003, http://www.nytime.com/2003/05/09/us/clinics-hold-more-embryos-than-had-been-thought.html?pagewanted=print (accessed April 29, 2014).
6. Curzer 2004, 545.

Understanding the Text

1. Deontology and utilitarianism are defined at the beginning of the piece. Where do you see each of those philosophies being applied throughout the reading? Identify one or two examples for each.
2. Based on this article alone, what is your understanding of how and why stem cells are used in medical research?
3. The author chose an interview format for this fictional piece; what effect does this format have on you as a reader? Why do you think Krimsky chose this format and is it successful in achieving his objective?

Reflection and Response

4. In the excerpt, Dr. Franklin asks Dr. Baum why German scientists are comfortable working with embryonic stem cells even though their very existence is based on the destruction of the embryo, which in Germany is illegal and considered immoral. What do you make of Dr. Baum's response to this question? Is it logical, hypocritical, or both? Explain your answer.
5. Dr. Baum poses two different questions: "Can we do good things through immoral acts? And should we use the outcome of bad acts to do good things?" (par. 10). Consider each of these questions individually and answer them for yourself based on your own convictions. Then, consider how the questions overlap – in what ways are they similar? What word choice makes them different?

Making Connections

6. The author begins by discussing the ethical implications of building upon the results of scientific experiments performed by Nazis. Research some of their experiments; what stands out to you? What are some arguments against using their results? What consequences, positive or negative, might their

continued use mean for science and how might those consequences apply to other sciences? Explain your answer.

7. Much of this reading is focused on the controversial use of embryonic stem cells and how they are obtained. After completing some additional research on the use of embryonic stem cells in scientific research, summarize an argument in favor of using them for scientific research and an argument against using them for scientific research. Then, develop three questions you have based on what you have learned.
8. Investigate how stem cell research is currently regulated in the United States and other countries like England, Germany, China, and South Africa. How do the regulations vary? What agencies are involved in the regulation? How might these differences impact the progression of scientific research reliant on stem cells?

All Too Human

Hillary Rosner

Hillary Rosner is a science and environment journalist who has written for multiple publications, including *Nature, Discover, National Geographic*, and *Popular Science*. She has both a master's degree in environmental science and a master of fine arts degree in creative writing. In this 2016 article from *Scientific American*, Rosner delves into the human desire for immortality, the fictional ways we've imagined we could accomplish it, and the reality of our scientific capabilities.

Countless books and films have imagined how humans might obtain immortality via science, whether it be by uploading our memories to a computer or becoming half-machine cyborgs. Here, Rosner tells us that real-world science is nowhere close to making these ideas possible, and it may never be. She argues, though, that won't stop people from dreaming about it. Her interrogation of the question embedded in the essay's title forces readers to step back and consider what living forever would really look like.

If We Could, Would We Want to Live Forever?

Recently, at a wedding reception, I polled some friends about immortality. Suppose you could upload your brain tomorrow and live forever as a human-machine hybrid, I asked an overeducated couple from San Francisco, parents of two young daughters. Would you do it? The husband, a forty-two-year-old MD-PhD, didn't hesitate before answering yes. His current research, he said, would likely bear fruit over the next several centuries, and he wanted to see what would come of it. "Plus, I want to see what the world is like 10,000 years from now." The wife, a thirty-nine-year-old with an art history doctorate, was also unequivocal. "No way," she said. "Death is part of life. I want to know what dying is like."

I wondered if his wife's decision might give the husband pause, but I diplomatically decided to drop it. Still, the whole thing was more than simply dinner party fodder. If you believe the claims of some futurists, we'll sooner or later need to grapple with these types of questions because, according to them, we are heading toward a postbiological world in which death is passé—or at least very much under our control.

The most well-imagined version of this transcendent° future is Ray Kurzweil's. In his 2005 best-selling book *The Singularity Is Near,* Kurzweil predicted that artificial intelligence would soon "encompass all human

Transcendent: beyond or above normal human experience/capability.

knowledge and proficiency." Nanoscale° brainscanning technology will ultimately enable "our gradual transfer of our intelligence, personality, and skills to the nonbiological portion of our intelligence." Meanwhile billions of nanobots inside our bodies will "destroy pathogens, correct DNA errors, eliminate toxins, and perform many other tasks to enhance our physical well-being. As a result, we will be able to live indefinitely without aging." These nanobots will create "virtual reality from within the nervous system." Increasingly, we will live in the virtual realm, which will be indistinguishable from that anemic universe we might call "real reality."

Based on progress in genetics, nanotechnology and robotics and on the exponential rate of technological change, Kurzweil set the date for the singularity—when nonbiological intelligence so far exceeds all human intelligence that there is "a profound and disruptive transformation in human capability"—at 2045. Today a handful of singulatarians still hold to that date, and progress in an aspect of artificial intelligence known as deep learning has only encouraged them.

Most scientists, however, think that any manifestation of our cyborg destiny is much, much farther away. Sebastian Seung, a professor at the Princeton Neuroscience Institute, has argued that uploading the brain may never be possible. Brains are made up of 100 billion neurons, connected by synapses; the entirety of those connections make up the connectome, which some neuroscientists believe holds the key to our identities. Even by Kurzweilian standards of technological progress, that is a whole lot of connections to map and upload. And the connectome might be only the beginning: neurons can also interact with one another outside of synapses, and such "extrasynaptic interactions" could turn out to be essential to brain function. If so, as Seung argued in his 2012 book *Connectome: How the Brain's Wiring Makes Us Who We Are*, a brain upload might also have to include not just every connection, or every neuron, but every atom. The computational power required for that, he wrote, "is completely out of the question unless your remote descendants survive for galactic timescales." 5

Still, the very possibility of a cyborg future, however remote or implausible, raises concerns important enough that legitimate philosophers are debating it in earnest. Even if our technology fails to achieve the full Kurzweilian vision, augmentation of our minds and our bodies may take us part of the way there—raising questions about what makes us human.

Nano: like nanobots or nanoscale; refers to something imperceptibly tiny.

Bruce Duncan of Terasem Movement Foundation poses with Bina48, a robot capable of engaging in conversation and whose thought process is built on that of a real person named Bina.
Kevin Murakami/Champlain College

I ask David Chalmers, a philosopher and codirector of the Center for Mind, Brain and Consciousness at New York University who has written about the best way to upload your brain to preserve your self-identity, whether he expects he will have the opportunity to live forever. Chalmers, who is fifty, says he doesn't think so—but that "absolutely these issues are going to become practical possibilities sometime in the next century or so."

Ronald Sandler, an environmental ethicist and chair of the department of philosophy and religion at Northeastern University, says talking about our cyborg future "puts a lot of issues in sharp relief. Thinking about the limit case can teach you about the near-term case."

And, of course, if there is even the remote possibility that those of us alive today might ultimately get to choose between death or immortality as a cyborg, maybe it's best to start mulling it over now. So putting aside the question of feasibility, it is worth pausing to consider more fundamental questions. Is it desirable? If my brain and my consciousness were uploaded into a cyborg, who would I be? Would I still love my family and friends? Would they still love me? Would I, ultimately, still be human?

10

One of the issues philosophers think about is how we treat one another. Would we still have the Golden Rule in a posthuman world? A few years ago Sandler coauthored a paper, "Transhumanism, Human Dignity, and Moral Status," arguing that "enhanced" humans would retain a moral obligation to regular humans. "Even if you become enhanced in some way, you still have to care about me," he tells me. Which seems hard to argue with—and harder still to believe would come to pass.

Other philosophers make a case for "moral enhancement"—using medical or biomedical means to give our principles an upgrade. If we're going to have massive intelligence and power at our disposal, we need to ensure Dr. Evil won't be at the controls. Our scientific knowledge "is beginning to enable us to directly affect the biological or physiological bases of human motivation, either through drugs, or through genetic selection or engineering, or by using external devices that affect the brain or the learning process," philosophers Julian Savulescu and Ingmar Persson wrote recently. "We could use these techniques to overcome the moral and psychological shortcomings that imperil the human species."

In an op-ed this past May in *The Washington Post* entitled "Soon We'll Use Science to Make People More Moral," James Hughes, a bioethicist and associate provost at the University of Massachusetts Boston, argued for moral enhancement, saying it needs to be voluntary rather than coercive. "With the aid of science, we will all be able to discover our own paths to technologically enabled happiness and virtue," wrote Hughes, who directs the Institute for Ethics and Emerging Technologies, a progressive transhumanist think tank. (For his part, Hughes, fifty-five, a former Buddhist monk, tells me that he would like to stay alive long enough to achieve enlightenment.)

There is also the question of how we might treat the planet. Living forever, in whatever capacity, would change our relationship not just to one another but to the world around us. Would it make us more or less concerned about the environment? Would the natural world be better or worse for it?

The singularity, Sandler points out to me, describes an end state. To get there would involve a huge amount of technological change, and "nothing changes our relationship with nature more quickly and robustly than technology." If we are at the point where we can upload human consciousness and move seamlessly between virtual and non-virtual reality, we will already be engineering nearly everything else in significant ways. "By the time the singularity would occur, our relationship with nature would be radically transformed already," Sandler says.

Although we would like to believe otherwise, in our current mere mortal state we remain hugely dependent on—and vulnerable to—natural systems. But in this future world, those dependencies would change. If we didn't need to breathe through lungs, why would we care about air pollution? If we didn't need to grow food, we would become fundamentally disconnected from the land around us. 15

Similarly, in a world where the real was indistinguishable from the virtual, we might derive equal benefit from digitally created nature as from the great outdoors. Our relationship to real nature would be altered. It would no longer be sensory, physical. That shift could have profound impacts on our brains, perhaps even the silicon versions. Research shows that interacting with nature affects us deeply—for the better. A connection to nature, even at an unconscious level, may be a fundamental quality of being human.

If our dependence on nature falls away, and our physical ability to commune with nature diminishes, then "the basis for environmental concern will shift much more strongly to these responsibilities to nature for its own sake," Sandler says. Our capacity for solving environmental problems—engineering the climate, say—will be beyond what we can imagine today. But will we still feel that nature has intrinsic value? If so, ecosystems might fare better. If not, other species and the ecosystems they would still rely on might be in trouble.

Our relationship to the environment also depends on the question of timescales. From a geologic perspective, the extinction crisis we are witnessing today might not matter. But it does matter from the timeline of a current human life. How might vastly extended life spans "change the perspective from which we ask questions and think about the nonhuman environment?" Sandler asks. "The timescales really matter to what reasonable answers are." Will we become more concerned about the environment because we will be around for so long? Or will we care less because we will take a broader, more geologic view?

"It's almost impossible to imagine what it will be like," Sandler says, "but we can know that the perspective will be very, very different."

Talk to experts about this stuff for long enough, and you fall down a rabbit hole; you find yourself having seemingly normal conversations about absurd things. "If there were something like an X-Men gene therapy, where they can shoot lasers out of their eyes or take over your mind," Hughes says to me at one point, then people who want those traits should have to complete special training and obtain a license. 20

"Are you using those examples to make a point, or are they actual things you believe are coming?" I ask.

"In terms of how much transhumanists talk about these things, most of us try not to freak out newbies too much," he replies obliquely. "But once you're past shock level 4, you can start talking about when we're all just nanobots."

When we're all just nanobots, what will we worry about? Angst, after all, is arguably one of our defining qualities as humans. Does immortality render angst obsolete? If I no longer had to stress about staying healthy, paying the bills, and how I'll support myself when I'm too old and frail to travel around writing articles, would I still be me? Or would I simply be a placid, overly contented . . . robot? For that matter, what would I daydream about? Would I lose my ambition, such as it is? I mean, if I live forever, surely that Great American Novel can wait until next century, right?

Would I still be me? Chalmers believes this "is going to become an extremely pressing practical, not just philosophical, question."

On a gut level, it seems implausible that I would remain myself if my brain was uploaded—even if, as Chalmers has prescribed, I did it neuron by neuron, staying conscious throughout, becoming gradually 1 percent silicon, then 5, then 10 and onward to 100. It's the old saw about Theseus's ship—replaced board by board with newer, stronger wood. Is it or isn't it the same afterward? If it's not the same, at what point does the balance tip?

"On a gut level, it seems implausible that I would remain myself if my brain was uploaded. . ."

"A big problem," Hughes says, "is you live long enough and you'll go through so many changes that there's no longer any meaning to you having lived longer. Am I really the same person I was when I was five? If I live for another 5,000 years, am I really the same as I am now? In the future, we will be able to share our memories, so there will be an erosion of the importance of personal identity and continuity." That sounds like kind of a drag.

Despite the singularity's utopian rhetoric, it carries a tinge of fatalism: this is the only route available to us; merge with machines or fade away—or worse. What if I don't want to become a cyborg? Kurzweil might say that it's only my currently flawed and limited biological brain that prevents me from seeing the true allure and potential of this future. And that the choices available to me—any type of body, any experience in virtual reality, limitless possibilities for creative expression, the chance to colonize space—will make my current biological existence seem almost comically trivial. And anyway, what's more fatalistic than certain death?

Nevertheless, I really like being human. I like knowing that I'm fundamentally made of the same stuff as all the other life on Earth. I'm even sort of attached to my human frailty. I like being warm and cuddly and not hard and indestructible like some action-film super-robot. I like the warm blood that runs through my veins, and I'm not sure I really want it replaced by nanobots.

Some ethicists argue that human happiness relies on the fact that our lives are fleeting, that we are vulnerable, interdependent creatures. How, in a human-machine future, would we find value and meaning in life?

"To me, the essence of being human is not our limitations . . . it's our ability to reach beyond our limitations," Kurzweil writes. It's an appealing point of view. Death has always fundamentally been one of those limitations, so perhaps reaching beyond death makes us deeply human? 30

But once we transcend it, I'm not convinced our humanity remains. Death itself doesn't define us, of course—all living things die—but our awareness and understanding of death, and our quest to make meaning of life in the interim, are surely part of the human spirit.

In Brief

Some scientists believe that one day technology will make it possible to achieve immortality by uploading our neural connections into robots' bodies; others believe this is impossible.

Regardless, legitimate philosophers are engaged in a debate over how such an eventuality would change our humanity.

Their dialogue is important because even if the "singularity" falls short, human augmentation and improvements may raise similar issues.

Understanding the Text

1. The "possibility of a cyborg future" appears to be "remote or implausible" (par. 6). Why? What barriers are there to humans uploading their brains to machines or replacing them entirely with machines?
2. Rosner discusses "moral enhancement" (par. 11) both as a possibility for future cyborgs, and also as an option for regular humans. How does she describe moral enhancement? Does she make an argument for or against it? Explain with support from the text.
3. Rosner goes beyond simply explaining how a society of cyborgs would change humanity. What questions does she have about environmental impact? Why is this important to consider?

Reflection and Response

4. Rosner poses several intriguing questions throughout her article. Choose one of her questions and write a response based on your own life experience and beliefs.
5. Much like "The Argument of Future Generations" (p. 63) in Chapter 1, this reading is more philosophical than scientific in nature. Especially because the real-world science is not capable of manifesting the ideas being proposed in Rosner's article, what is the value of asking philosophical questions about immortality? How does Rosner's piece model critical inquiry for this topic and for others?

Making Connections

6. The replicants in *Blade Runner*, the Borg in *Star Trek*, Vision in the *Avengers* series, the Terminator, the AI of *Ex Machina* and *Her*, and even the hybrid characters in the Netflix series *Altered Carbon* represent just a few of the many sci-fi visions that have embodied what it means to be "human." Consider some of the reasons for desiring immortality that Rosner provides in her article. Compare the desires played out on-screen in one or more of these stories – what connections do you see? Are the outcomes of humankind pursuing these desires largely positive, negative, or neutral?
7. In his book *To Be a Machine: Adventures among Cyborgs, Utopians, Hackers, and the Futurists Solving the Modest Problem of Death*, Mark O'Connell claims that transhumanists believe "we can and should eradicate ageing as a cause of death; that we can and should use technology to augment our bodies and our minds; that we can and should merge with machines, remaking ourselves, finally, in the image of our own higher ideals." Investigate the transhumanist philosophy and human enhancement. What are the beliefs and goals behind these movements, and what real scientific progress is being made toward meeting their goals?
8. Books like Luke Dormehl's *Thinking Machines: The Quest for Artificial Intelligence and Where It's Taking Us Next*, and John Brockman's *What to Think about Machines That Think* explore what scientists and engineers themselves are saying about the future of AI. Refer to either Dormehl's or Brockman's book, or choose a similar text to consider. What concerns do scientists and engineers have? What hopes? How might these differ from those of the average person?

The Horror Story That Haunts Science: Two Hundred Years Later, *Frankenstein* Still Shocks and Inspires

Jon Cohen

Jon Cohen has been a staff writer at *Science* magazine for nearly thirty years. During his career, he has also written for other publications including *Technology Review*, the *New Yorker*, and *Smithsonian*. He is the author of four books, and he has received multiple awards and honors for his work both as a journalist and documentary filmmaker.

In this 2018 article from *Science*, Cohen reminds readers that the importance of Mary Shelley's *Frankenstein* cannot be overstated even in modern scientific circles. Considered by many to be the forerunner of the science fiction genre, her novel described the ill-conceived attempts by Victor Frankenstein to overcome death. His failure to foresee the consequences of such an unnatural creation has become the ultimate cautionary tale for scientists in the last two centuries. But why? Here Cohen provides an in-depth look at the symbiotic relationship between the novel and real-world science, both past and present.

On August 1, 1790, a precocious student named Victor Frankenstein submitted a radical proposal to an ethical panel at the University of Ingolstadt in Bavaria. Under the title "Electro-chemical Mechanisms of Animation," Frankenstein explained how he wanted to "reverse the processes of death" by collecting "a large variety of human anatomical specimens" and putting them together to try and "restore life where it has been lost."

Frankenstein assured the institutional review board (IRB) that he had the highest ethical standards. "If I do succeed in fully animating a human or human-like creature, I will provide the creature with information about the study and allow it, if it is capable, to choose whether or not to participate further in continued observation and study," noted the budding scientist. If the creature had "diminished capacity," Frankenstein promised to bring in a third party to act in its interest and treat "the being" in accordance with recognized standards.

Of course no such proposal ever went to bioethicists at the University of Ingolstadt, where the fictional Frankenstein created his monster. In 1790, even a real Frankenstein would have faced no ethical reviews. But the proposal does exist in a 2014 paper, which speculates about whether the *Frankenstein* story would have had a happier ending if twenty-first-century safeguards had existed two centuries ago. It is one of many riffs on the novel to be found in biomedical literature. In conceiving her story, Mary Shelley was influenced by the nascent medical science of the day and by early experiments on electricity. In return, *Frankenstein* has haunted science ever since.

First published anonymously in 1818, the book and subsequent films and plays have become what Jon Turney, author of the book *Frankenstein's Footsteps: Science, Genetics and Popular Culture,* calls "the governing myth of modern biology": a cautionary tale of scientific hubris. And as with all long-lasting myths, it is not one myth, but many, as a search for "Frankenstein" in the PubMed database—the main catalog of life sciences papers—makes clear. Scientific literature, like the popular press, is rife with references to Frankenfood, Frankencells, Frankenlaws, Frankenswine, and Frankendrugs—most of them supposedly monstrous creations. Other papers explicitly mentioning *Frankenstein*—there are more than 250 of them—analyze the science behind the novel or even, in a twist that can be downright bizarre, draw inspiration from it.

Several reports in psychological journals delve into the state of mind of its author when she first imagined the tale during the summer of 1816. Then Mary Wollstonecraft Godwin, she was visiting the poet Lord Byron at Villa Diodati, a mansion he had rented on the shores of Lake Geneva in Switzerland. She was eighteen, accompanying her married lover, the poet Percy Bysshe Shelley. Her stepsister, Claire Clairmont, was there, as was Byron's live-in doctor, John William Polidori. It was the "year without a summer," a climatic anomaly caused by the eruption of Mount Tambora in the Dutch East Indies, and endless rain and gray skies kept the guests cooped up. Byron suggested as a party game that they each write a ghost story.

There was plenty to unsettle Mary's fertile mind. Mary and Percy had a 6-month-old baby together, but had lost another baby a year earlier. Mary's own mother had died of puerperal sepsis eleven days after giving birth to her fame-bound daughter. Percy, as a 2013 paper in *Progress in Brain Research* recounts, had been booted from the University of Oxford in the United Kingdom for "extolling the virtues of atheism" and was a believer in "free love." Another paper, in a 2015 issue of *The Journal of Analytical Psychology,* suggests that Percy, Mary, and Claire had previously formed "a ménage à trois of sorts."

The Journal of Analytical Psychology paper's author, Ronald Britton, a prominent psychoanalyst, links these tensions and griefs to the daydream in which Mary Shelley first envisioned Frankenstein's monster—"the spectre which had haunted my midnight pillow," as she later put it. The "background facts to her nightmare," Britton writes, invoking Freud, "opened a door to unconscious phantasies of a dreadful scene of childbirth." He adds that after losing her first child in 1815, Shelley wrote in her journal that she dreamed about the baby coming back to life. "I thought that if I could bestow animation upon lifeless matter, I might, in process of time, renew life where death had apparently devoted the body to corruption," she wrote the year before imagining *Frankenstein*.

More horrors were to follow for Shelley after she completed the novel. She married Percy after his first wife's suicide, only to lose him 6 years later when he drowned in a sailing accident. But she called on science, not psychology, in explaining how she "came to think of, and dilate upon, so very hideous an idea" at eighteen years of age. Among the influences she cites in a preface to an 1831 edition of her novel is Luigi Galvani, who in 1780 found that an electrical charge could make a dead frog's legs twitch. It was Percy who may have acquainted her with galvanism, which Frankenstein explicitly mentions as the key to reanimation in the 1831 edition. As a boy, the poet "had dabbled with electricity (on his sister's sores and the family cat)," another study in Progress in Brain Research notes.

Many a paper has attempted to parse other ways in which the science of the day influenced Shelley's tale. A 2016 essay in *Nature* by a U.K. biographer noted that her novelist father was friends with electrochemist Humphry Davy and with William Nicholson, a co-discoverer of electrolysis, the technique of triggering chemical reactions using electricity. Several accounts point to the influence of Byron's physician, Polidori (who later poisoned himself with prussic acid), and his discussions of

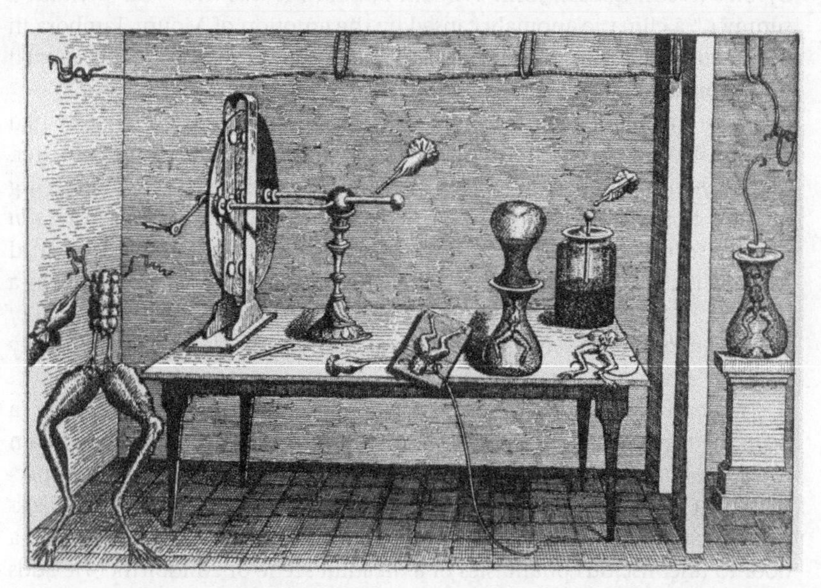

A 1791 engraving of Luigi Galvani's experiments to animate a frog's legs using electricity.
Time & Life Pictures/Getty Images

experiments on spontaneous generation by Erasmus Darwin, grandfather of Charles. A 2004 paper in the *Journal of Clinical Neurophysiology* that reviews the "electrophysiological undercurrents for Dr. Frankenstein" notes that Shelley could not have missed the widely discussed work of Giovanni Aldini, Galvani's nephew, who in 1803 zapped the heads of decapitated criminals in an attempt to reanimate them; he imagined this could be used to resuscitate people who had drowned or suffocated and possibly to help the insane.

Over time, the influence ran from the novel back to science. "From Frankenstein to the Pacemaker," in *IEEE Engineering in Medicine and Biology Magazine*, tells how eight-year-old Earl Bakken in 1932 saw the famous *Frankenstein* movie starring Boris Karloff, which "sparked Bakken's interest in combining electricity and medicine." Bakken would later found Medtronic, develop the first transistorized cardiac pacemaker, and open a museum devoted to electricity in the life sciences that's housed in a Gothic Revival style mansion in Minneapolis, Minnesota. Neighborhood kids call it Frankenstein's castle. 10

Indeed, many scientific studies proudly reference *Frankenstein,* mainly because they combine disparate parts to create a novel entity that the researchers present as delightfully chimeric. A milk sugar enzyme fused with a carrier protein. An atlas of the head and neck to guide radiotherapy, created by merging views from different patients. A face recognition study that swapped the eyes, noses, and mouths of former President George W. Bush and former U.S. Secretary of State Colin Powell. A "Frankenrig" used to create 3D animations, made by mixing and matching bones from different skeletons.

In perhaps the strangest embrace of the *Frankenstein* label, a 2013 article in *Surgical Neurology International* proposes recreating Aldini's electrifying head experiments. The authors of "HEAVEN: The Frankenstein effect," note that Aldini ultimately aimed to transplant a human head, using electricity to spark it back into awareness. That's just what the authors have in mind for their project, the head anastomosis venture (HEAVEN). "On the whole, in the face of clear commitment, HEAVEN could bear fruit within a couple of years," they write. (Many scientists have called the project unfeasible and unethical, but last November, two of the coauthors announced to the media that they had performed a head transplant on a human corpse and soon planned to publish details.)

But by far the bulk of the scientific literature hand-wrings, ponders, and philosophizes about the most familiar form of the *Frankenstein* myth, which Shelley flicked at in her "Modern Prometheus" subtitle: the idea that mad scientists playing God the creator will cause the entire human species to suffer eternal punishment for their trespasses and hubris.

"Mary Shelley, Frankenstein, and The Dark Side of Medical Science," a 2014 essay published in the charmingly incongruous *Transactions of the American Clinical and Climatological Association,* ticks off a diverse list of recent experiments that have drawn the "Franken-" label: the cloning of Dolly the sheep, the engineering of a highly lethal H5N1 bird influenza that could more easily infect mammals, the synthesizing of an entire bacterial genome. Other triggers of *Frankenstein*-ish fears have included in vitro fertilization, proposals to transplant pig organs into humans, and tomatoes endowed with genes from fish to make them freeze-tolerant.

J. Craig Venter, a pioneer in genomics based in San Diego, California, has been called a Frankenstein for his effort to create artificial bacteria with the smallest possible genomes. Still, he's a fan of Shelley's tale. "I think she's had more influence with that one book than most authors in history," says Venter, who owns a first edition. "It affects a lot of people's thinking and fear because it represents this fundamental of 'You don't mess with Mother Nature and you don't mess with life because God will strike you down.'"

"Obviously, I don't buy into that theme," he adds.

> "The *Frankenstein* myth endures, he says, because 'fear is easy to sell'—even when unwarranted."

The *Frankenstein* myth endures, he says, because "fear is easy to sell"—even when unwarranted. "Most people have a fear of what they don't understand," he says. "Synthetic cells are pretty complicated and putting a new gene into corn sounds scary." But by throwing around labels like Frankenfood and Frankencells to rally the public against potentially valuable innovations, he says, the "fear-based community will potentially do more damage to humanity than the things they fear."

Unlike the Frankenstein character, who initially didn't consider how his work might go wrong, Venter says he recognizes that editing and rewriting genomes could "contaminate the world" and cause unintended harm. "I think we need to be very smart about when we do it and how we do it," he says. He thinks Shelley "would highly appreciate" his work.

Henk van den Belt, a philosopher and ethicist at Wageningen University in the Netherlands who wrote a paper about *Frankenstein* and synthetic biology, applauds Venter for fighting back against the Frankenslur. "Very often scientists are afraid to take this position, but I think it's better to be defiant," Van den Belt says. "Rhetoricians and journalists can accuse people of playing Frankenstein, but it's a little too easy. If scientists challenge this phrase, it will have less impact."

Shelley of course couldn't have imagined any of this hubbub, and indeed her tale has been wildly distorted in the popular imagination over the past two centuries. Frankenstein's aim was not to rule the world à la Dr. Evil, but "to banish disease from the human frame and render man invulnerable to any but a violent death." And Britton, the psychoanalyst, notes that the creature did not begin life as a monster; he only went on a killing spree because he sought love and happiness but was abhorred by his creator, who referred to him as "devil," "fiend," "abortion," "daemon," "vile insect," and other terms that would have made an IRB contact the Office for Human Research Protections. "I was benevolent and good, misery made me a fiend," Frankenstein's creation said. "[I]mpotent envy and bitter indignation filled me with an insatiable thirst for vengeance." 20

A dental radiologist of all people published an insightful two-part essay in *The Journal of the Royal Society of Medicine* in 1994 that underscores what some argue is the real moral of the book: not the danger of scientists violating the natural order, but the dire fate that awaits creators who fail to care for their creations. "Read the book and weep for those we have rejected, and fear for what revenge they will exact, but shed no tears for Frankenstein," the essay advises, referring to the doctor. "Those who think, in ignorance of the book, that his is the name of the Monster are in reality more correct than not."

Understanding the Text

1. Cohen pulls from the experiences of a variety of people to support his claims in this article. Choose two pieces of evidence from the article that were particularly effective in explaining or illuminating a topic or claim.
2. According to Cohen, what personal and scientific events may have influenced Shelley's novel? What is the purpose of Cohen's article and who do you believe his audience to be?

Reflection and Response

3. *Frankenstein* has long been seen as a cautionary tale about what happens when humankind exceeds the limits of its own knowledge. From this article, choose one or two examples of a science or technology that you feel may be going too far. What have you learned that gives you this impression? What questions do you have for further research to confirm or refute this initial impression?
4. The article begins with an alternative version of *Frankenstein* wherein Dr. Frankenstein proposes his experiment to an Institutional Review Board (IRB) before he moves forward. Today, IRB approval is required for nearly all university research. Do you think this requirement makes sense, or might it add barriers to innovative research? Explain.

Making Connections

5. In the article, Dr. J. Craig Venter draws a contrast between himself and Dr. Frankenstein, arguing that "he recognizes that editing and rewriting genomes could 'contaminate the world' and cause unintended harm" (par. 17). Consider "The CRISPR Pioneers" (p. 82), or another reading from earlier in this chapter, which provides multiple examples of scientists considering the ramifications of their work. Is the stigma created by Victor Frankenstein's careless approach to science still appropriate given what you have learned so far? Explain your answer using examples from the other readings.
6. Many writers have been inspired by science, and vice versa. In 2012 the Smithsonian developed a list of ten scientific inventions that were inspired by literature. Some of those inventions included the submarine, which was inspired by Jules Verne's *Twenty Thousand Leagues Under the Sea*, and the cell phone, which was inspired by the original *Star Trek* communicator of the 1960s. Investigate other real-world scientific inventions that were inspired by science fiction.
7. This article is filled with references to outside sources that illustrate *Frankenstein*'s connection to science. Mine the sources of this article (including print sources and actual people) and choose two to investigate further.

From *Super Sad True Love Story*

Gary Shteyngart

Gary Shteyngart is a novelist who has published five books and won multiple awards, including the Stephen Crane for First Fiction, the National Jewish Book Award for Fiction, and the Bollinger Everyman Wodehouse Award. His work has been published in the *New York Times Magazine*, the *New Yorker*, and *GQ*, among others.

In this excerpt from the beginning of Shteyngart's novel, *Super Sad True Love Story* (2010), we meet Lenny Abramov via one of his diary entries. Lenny is an average-looking, middle-aged man living in a dystopic America. He also happens to work as a salesman for a company that claims to help people to live forever. Similar to the premises we've seen explored in our earlier readings (e.g., "All Too Human" and "The Horror Story That Haunts Science"), here, Lenny begins to consider the fragility of natural human life and the scientific possibilities for transcending it.

Do Not Go Gentle

FROM THE DIARIES OF LENNY ABRAMOV

JUNE 1
Rome—New York

Dearest Diary,

Today I've made a major decision: *I am never going to die.* Others will die around me. They will be nullified. Nothing of their personality will remain. The light switch will be turned off. Their lives, their entirety, will be marked by glossy marble headstones bearing false summations ("her star shone brightly," "never to be forgotten," "he liked jazz"), and then these too will be lost in a coastal flood or get hacked to pieces by some genetically modified future-turkey.

Don't let them tell you life's a journey. A journey is when you end up *some*where. When I take the number 6 train to see my social worker, that's a journey. When I beg the pilot of this rickety United-Continental Deltamerican plane currently trembling its way across the Atlantic to turn around and head straight back to Rome and into Eunice Park's fickle arms, *that's* a journey.

But wait. There's more, isn't there? There's our legacy. We don't die because our progeny lives on! The ritual passing of the DNA, Mama's corkscrew curls, his granddaddy's lower lip, *ah buh-lieve thuh chil'ren ah our future*. I'm quoting here from "The Greatest Love of All," by 1980s pop diva Whitney Houston, track nine of her eponymous first LP.

Utter nonsense. The children are our future only in the most narrow, transitive sense. They are our future until they too perish. The song's next line, "Teach them well and let them lead the way," encourages an adult's relinquishing of selfhood in favor of future generations. The phrase "I live for my kids," for example, is tantamount to admitting that one will be dead shortly and that one's life, for all practical purposes, is already over. "I'm gradually dying for my kids" would be more accurate.

"The children are our future only in the most narrow, transitive sense. They are our future until they too perish."

But what *ah* our *chil'ren*? Lovely and fresh in their youth; blind to mortality; rolling around, Eunice Park-like, in the tall grass with their alabaster legs; fawns, sweet fawns, all of them, gleaming in their dreamy plasticity, at one with the outwardly simple nature of their world.

And then, a brief almost-century later: drooling on some poor Mexican nursemaid in an Arizona hospice.

Nullified. Did you know that each peaceful, natural death at age eighty-one is a tragedy without compare? Every day people, individuals—*Americans*, if that makes it more urgent for you—fall facedown on the battlefield, never to get up again. Never to exist again.

These are complex personalities, their cerebral cortexes shimmering with floating worlds, universes that would have floored our sheepherding, fig-eating, analog ancestors. These folks are minor deities, vessels of love, life-givers, unsung geniuses, gods of the forge getting up at six-fifteen in the morning to fire up the coffeemaker, mouthing silent prayers that they will live to see the next day and the one after that and then Sarah's graduation and then . . .

Nullified.

But not me, dear diary. Lucky diary. Undeserving diary. From this day forward you will travel on the greatest adventure yet undertaken by a nervous, average man sixty-nine inches in height, 160 pounds in heft, with a slightly dangerous body mass index of 23.9. Why "from this day forward"? Because yesterday I met Eunice Park, and she will sustain me through forever. Take a long look at me, diary. What do you see? A slight man with a gray, sunken battleship of a face, curious wet eyes, a giant gleaming forehead on which a dozen cavemen could have painted something nice, a sickle of a nose perched atop a tiny puckered mouth, and from the back, a growing bald spot whose shape perfectly replicates the great state of Ohio, with its capital city, Columbus, marked by a deep-brown mole. *Slight.* Slightness is my curse in every sense. A so-so body in a world where only an incredible one will do. A body at the chronological age of thirty-nine already racked with too much LDL cholesterol, too much ACTH hormone, too much of everything that dooms the heart, sunders the liver, explodes all hope. A week ago, before Eunice gave me reason to live, you wouldn't have noticed me, diary. A week ago, I did not exist. A week ago, at a restaurant in Turin, I approached a potential client, a classically attractive High Net Worth Individual. He looked up from his wintry *bollito misto,* looked right past me, looked back down at the boiled lovemaking of his seven meats and seven vegetable sauces, looked back up, looked right past me again—it is clear that for a member of upper society to even remotely notice me I must first fire a flaming arrow into a dancing moose or be kicked in the testicles by a head of state.

And yet Lenny Abramov, your humble diarist, your small nonentity, will live forever. The technology is almost here. As the Life Lovers Outreach Coordinator (Grade G) of the Post-Human Services division of the Staatling-Wapachung Corporation, I will be the first to partake of it. I just have to be good and I have to believe in myself. I just have to stay off the trans fats and the hooch. I just have to drink plenty of green tea and alkalinized water and submit my genome to the right people. I will need to re-grow my melting liver, replace the entire circulatory system with "smart blood," and find someplace safe and warm (but not too warm) to while away the angry seasons and the holocausts. And when the earth expires, as it surely must, I will leave it for a new earth, greener still but with fewer allergens; and in the flowering of my own intelligence some 1032 years hence, when our universe decides to fold in on itself, my personality will jump through a black hole and surf into a dimension of unthinkable wonders, where the things that sustained me on Earth 1.0—*tortelli lucchese,* pistachio ice cream, the early works of the Velvet Underground, smooth, tanned skin pulled over the soft Baroque architecture of twentysomething buttocks—will seem as laughable and infantile as building blocks, baby formula, a game of "Simon says *do this.*"

That's right: I am never going to die, *caro diario.* Never, never, never, never. And you can go to hell for doubting me.

Understanding the Text

1. Lenny describes himself in great detail. What is your impression of both his physical appearance and his personality? How do you think he feels about *himself?*
2. Lenny works for "the Post-Human Services division of the Staatling-Wapachung Corporation" (par. 11). What does the wording, "Post-Human Services," imply? Consider your own impression of the term and information the excerpt provides. Explain your answer.

3. Lenny ends the diary entry with "and you can go to hell for doubting me." He addresses the diary itself more than once, so who is he talking to? Why do you think he says this?

Reflection and Response

4. The very first lines describe his "major decision" to "never die." Are there any circumstances that would lead you to consider living forever? Explain your answer.
5. Lenny explains that we are not really living for our kids; rather, we are dying for them. What does he mean by this? Do you agree or disagree with his assessment?
6. The word *nullified* is repeated three times. Why do you think Shteyngart chose that word and what purpose does this repetition serve?

Making Connections

7. Compare this excerpt to Hillary Rosner's "All Too Human" (p. 118). What connections do you see between the two? How do Lenny's descriptions of humankind illustrate what it means to be "all too human?"
8. Dr. Frankenstein, a manifestation of Mary Shelley's own desires, sought to "reverse the processes of death." If we follow that idea, its natural next step would mean humans could avoid death and live forever. Although Lenny is still very much alive, he is also seeking to avoid death and to live forever. Research other science fiction literature where the character(s) wanted to live forever. What did they do to accomplish it? What were their motivations?
9. Research some modern science and technology, such as pacemakers and dialysis, that we already use to preserve and extend human life. Are these really small steps toward immortality? What motivations do we have for using them and how do they differ, if at all, from the motivations discussed in Shteyngart's and Rosner's pieces?

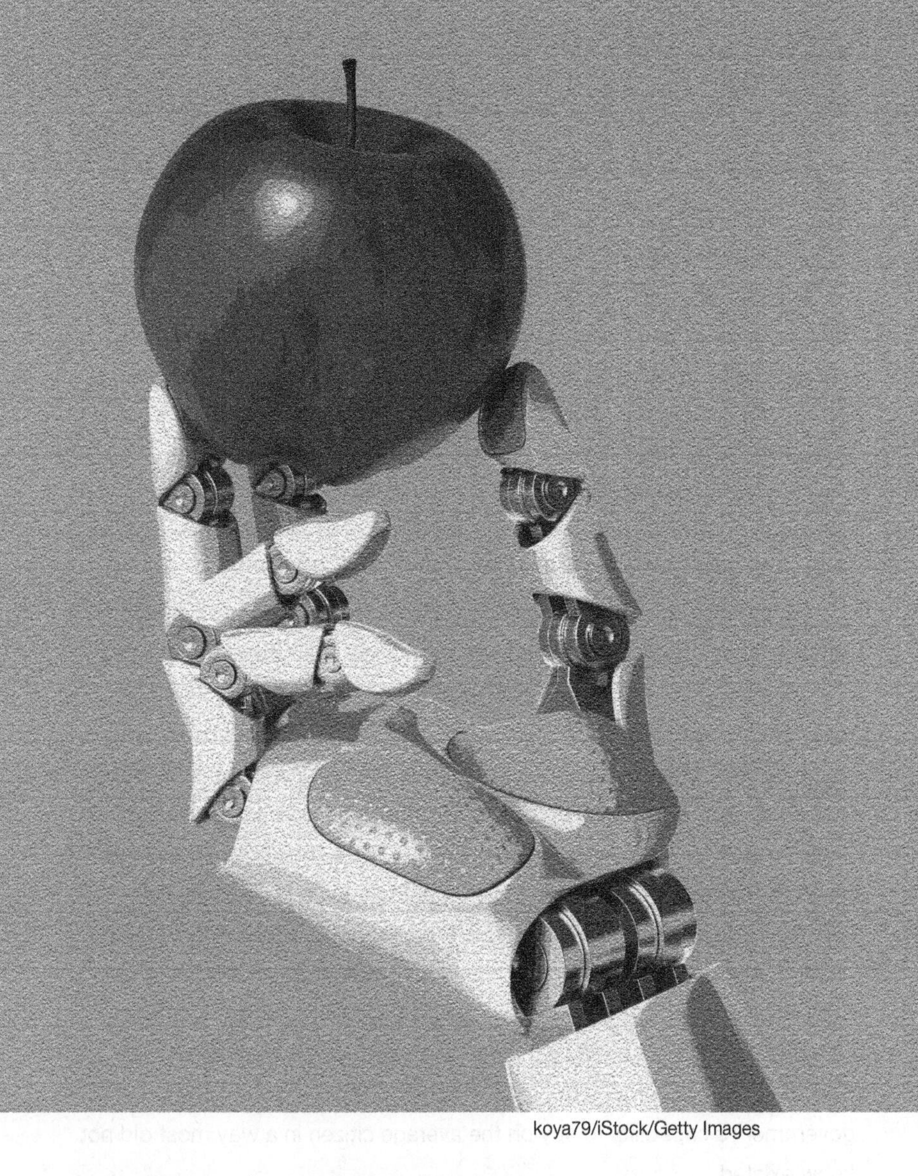

koya79/iStock/Getty Images

3 Have You Been Spied On Today?

Take a moment and look around. What types of technology surround you? Most of us are often within arm's reach of a smartphone or a computer. But how many of us are aware of technology's capability to spy on us? In today's world, technology is routinely used to garner information about individuals, groups of people, and even large companies — both legally and illegally. Who, then, is doing the spying? People are often surprised to learn our own government routinely taps into social media, smartphones, and more to gather information for a variety of reasons. When did the government begin spying on the average citizen in such a ubiquitous way, and why aren't more of us aware or concerned about it?

After the attacks on September 11, 2001, Congress passed the Patriot Act, which legally expanded the government's surveillance program in an attempt to help combat future terrorist attacks. Despite good intentions, many critics have argued the Patriot Act violates the Fourth Amendment, which protects citizens against unlawful searches and seizures. Still, the Patriot Act has remained enacted under multiple presidents and has now been in place for over a decade, and since 2001, technological advances have only made it easier for the government to spy on its citizens. For example, Edward Snowden, a famous — or infamous, depending on your point of view — former government employee exposed how the National Security Agency (NSA) was illegally gathering data on random American citizens via cell phones. While the country is divided on whether or not Snowden is a hero or a traitor, the fact remains that he revealed the government's capability to spy on the average citizen in a way most did not know existed.

This type of surveillance is not limited to the United States, however. Nicole Perlroth, author of "Governments Turn to Commercial Spyware to Intimidate Dissidents," exposes how a human rights activist from the United Arab Emirates was targeted and arrested by his government. Because spyware is inexpensive and accessible, it was easy for his government to

photo: koya79/iStock/Getty Images

turn the technology in his own home against him. But the government is not the only entity capable of spying. Scott Lucas explains how Facebook data was mined for political purposes in his article "Why Cambridge Analytica Matters." Lucas explains how social media and the data it houses was weaponized and used to manipulate public opinion, possibly affecting the last presidential election. These are extreme scenarios, so we will also consider some of the more subtle ways technology acts as an often porous gatekeeper, unknowingly (and knowingly) leaking personal information.

This chapter explores and investigates how technology leaves the average citizen vulnerable to surveillance and privacy leaks. So what does this mean for our day-to-day lives? William Eyre, author of "Surveillance Today," explains how something as simple as a DVR, TiVo, or a cable satellite provider stores information about what shows a customer watches in his or her home. Seems benign enough, right? It's just television, but what happens when the company sells that data? Is that *still* benign? Should *that* be legal? What about apps like Angry Birds or Snapchat? Stuart Sumner, author of *You: For Sale* explains how these apps put their users' information at risk — sometimes because the technology fails and unintentionally leaks user information and sometimes because the government takes advantage of these leaks. Oftentimes it's only metadata, not detailed content, explains Steven Aftergood in "Privacy and the Imperative of Open Government," but still, he encourages people to care, yet many do not.

Today, people will repeat some version of this old adage: "I have nothing to hide, so it is fine if I'm spied on." Those opposed to quasi-legal surveillance, however, will quote Benjamin Franklin: "Those who would give up essential Liberty, to purchase a little temporary Safety, deserve neither Liberty nor Safety." Both sides posit solid arguments defending each stance. Despite your personal take on the issue, the fact remains that technology is enhancing the ability to spy in new ways every day, and these enhancements grow every day. Kurt Vonnegut's short story "Harrison Bergeron" takes these capabilities to a hyperbolic level, using fiction as a cautionary tale for surveillance possibilities in the future, but Vonnegut's ideas about government control remain relevant today. Facebook, a social media site that has over one billion users, is still banned in some countries according to Paul Mozur, Mark Scott, and Mike Isaac. Their article, "Facebook Faces a New World as Officials Rein in a Wild Web," explains that a Facebook ban is one way some governments maintain control over their citizens.

Just as governments try to control and surveil citizens, so do companies try to exert control over their employees. The chapter closes by examining how technology, specifically AI, is entering the workplace in "Smile, You're on Camera." AI presents the opportunity for companies to become more efficient, but at what cost? At what risk? As you read this chapter, think about the benefits, consequences, and possibilities technology is allowing with respect to spying. How much privacy are you willing to give up?

Surveillance Today; from *Real ID Act: Privacy and Government Surveillance*

William Eyre

William Eyre spent twenty years running and owning an IT-security consulting company. After earning a PhD from Purdue University, where he studied international terrorism, biometrics, cryptography, and computer forensics, complete with a dissertation in information security, William Eyre went on to work for the U.S. government.

His book, *The Real ID Act: Privacy and Government Surveillance* (2011), details the expansive spying capabilities technology has produced, allowing the government to tap into many day-to-day technologies with the purpose of procuring information about all of us. In the following excerpt, from the first chapter of his book, Eyre illustrates how seemingly benign, everyday technologies can track and log data about average citizens.

A Day in the Life

It's an average day. Winston and Julia Smith live in a high-rise condominium in Chicago and start the day by tuning to the morning news. Their television is fed using a cable box, TiVo or satellite connection and the couple watches the program on a High Definition Television (HDTV). The television signal provider logs and stores the information regarding the channels to which the Smiths tune at any given time, down to the second, and allows the provider the ability to record in a computer log file those preferences. Additionally, if the individual has a Digital Video Recorder, the time and duration, as well as the content of the show being viewed, are also known to TiVo or whatever company administers the DVR software updates and management (Charny, 2004).

DVRs conduct two-way communication with the entertainment content provider, gathering information on users' viewing choices, ostensibly for the purpose of giving entertainment consumers other, similar or related viewing choices. In order for this goal of predicting users' preferences to be achieved, the shows the consumers watch must be known to the preference prediction algorithm on a computer. This algorithm determines recommendations to the users based on that user's past viewing. This recommendation scheme also rests on some assumptions. Mainly that the targeted individual watches and enjoys the shows that appear on his/her screen. In cases in which the consumer leaves the power to the television on when out of the room or house, or uses the television to entertain guests, the assumptions start to break down.

However, in its most basic form, the "recommendations for future viewing" feature is implemented using an algorithm which determines the genre and subject matter of shows the users appear to like. At the algorithm's heart is the "rational choice" model in which it is assumed that users will only view the shows they like and not select and view shows they do not like (such as if they were assigned to watch a show for school or an influential person in the targeted consumer's sphere asked the consumer to watch a certain show). The algorithm then fits the users' past selections to a set of preference scoring criteria, picking other shows and/or movies matching those scoring criteria and recommending the resultant list to the viewer. This process is analogous to the manner in which Amazon makes recommendations to its customers regarding books the targeted consumer might prefer ("Million Dollar Netflix," 2006).

This information has commercial value and is stored indefinitely. It also has value in that it can give insight into individuals' interests. This viewing information provides data for content programmers and the types of organizations possessing the requisite motive and resources for understanding the psychology of the viewer.

If Winston or Julia were to connect a personal computer to the HDTV itself, with the HDTV acting as the monitor (screen) for the computer, the Smiths could perform all of their computer functions they would normally perform using the HDTV display. The Smiths could also connect their webcam in this configuration and the HDTV would provide higher definition fidelity when communicating with people at the other end of an internet connection for a chat session. The generally unintended consequence of communicating according to the parameters described in this scenario involves the fact that everything displayed on the HDTV acting as the computer monitor may be transmitted as data to the other end of an HDTV connection as a separate data stream of which the Smiths are unaware. This data stream would be in addition to the data communication with the other end of the internet connection. The communications would be transmitted easily as two separate connections. These connections could be to two essentially dissimilar places, as the processor in the HDTV (the computer that converts HD digital signal to build the picture) by definition has enough processing power to render the picture itself into a digital signal. This signal could then itself be sent to wherever it is told.

In the very near future, the Smiths will not need a webcam in order to transmit the goings on in the field of view of their television. Apple holds a patent which describes a television screen with sensors interlaced amongst the pixels which compose the display portion of the TV. These sensors interpret the photonic input they receive to allow the TV itself to act as a camera. To zoom and pull out, the array selects smaller or larger subsets of sensors which provide variable focal lengths. This serves the purpose of the TV having the capability to transmit close-ups and wide angle views (Fox, 2006).

One or both of the Smiths might then sign onto the internet and check e-mail. If one of the Smiths visits a government website, under a recent United States Department of Homeland Security (DHS) Cyber-Initiative, established by classified presidential order, any communication that traverses a federal government network may be recorded by the National Security Agency (NSA) or DHS (Nakashima, 2008a). The reason given for recording visits by American citizens seeking information from government websites is in order that the United States is protected from terrorist cyberattacks. All visits to government sites (i.e., IRS, CDC, DoJ, etc.) are recorded and NSA monitors and records all communications including internet, voice and e-mail (Bamford, 2008). AT&T also monitors the contents of internet communications in real time (Singel, 2007) and could easily record all the traffic that crosses their backbone.

In addition to any monitoring and recording of web use and e-mail traffic by the government and the backbone provider, all visited web addresses and e-mails are recorded in the ISP's database. Each click on an ad or link proffers a new piece of information for the ISP's database. The addresses of websites visited as well as the content of those sites are also, in the default configuration of the most used web browsers, stored on the user's hard disk. The reason for this storage is that the browser will then speed up the display of data (text and graphics) from the "cache" ("Deleting Web Browser," 2006). It also leaves a permanent, barring user intervention, record of everything the user has "seen." (Certain assumptions are necessary: that the user did not click on a link which maliciously redirected them to a site not described by the link, and that those links which were clicked correctly offered sites and pages that the user then actually viewed.)

In the Smiths' high-rise condo's hallway there is a camera. When they leave their condo, their images are recorded by a camera in the hallway. When they have visitors, the visitors are photographed entering and leaving the Smiths' condo. The elevator has camera. As well the elevator lobby, main lobby by the guard's station and doors to the outside are under video surveillance, as is the parking garage and the garage's entrance.

Winston works in the suburbs and drives to work from the city. Julia works in the city's center ("the Loop" in Chicago's case) and takes mass transit.

For the transit system to charge the correct fare, the customer's entrance and exit points from the system must be known. When a transit customer boards a train at a certain platform, the customer uses the Smart Card and the location and time is transmitted to and recorded in the database. Smart Cards communicate through the distributed Smart Card sensor network (i.e. the entrances and exits to the transit platforms at which the contactless cards are presented.)

The system processes the information on the entry and is performed in order to automatically deduct the amount of the travel from an account (card.) The account must be filled with money and this can be accomplished using a credit card. The ostensible purpose of the Smart Cards is to save money for the transit system and eliminate the fraud involved with people passing transfers to each other in order to get free rides (Godfrey, 2008).

Once the credit card is used for the purpose of payment, the transit card number and the credit card number are linked. Log entries must be made and kept of all transactions at each stage ("SmarTrip Questions," n.d.), and therefore all the data associated with an individual's transit system travels are known forever and can be mined or viewed any time thereafter ("Intelligent Transportation Systems," 2005).

Data mining is the technique of using databases as input for algorithms that search for patterns and identify characteristics and generators of patterns in which the data analysts' end users are interested ("Data Mining with MicroStrategy," n.d.).

In Winston's case, that of a commuter driving to work, the fact that he owns a car is registered with his state's motor vehicle bureau. The fact that the vehicle is registered can, on its own, be considered a minor issue in terms of surveillance. Vehicle registration has been required for many years for purposes of taxation. Massachusetts was the first state to issue registration tags in the United States in 1903 (Tortora, 1998) and these tags are basically tax receipts and therefore the tags show credit for the proper vehicle registrant. For surveillance purposes, the vehicle's plate number can be associated with an individual owner. If Winston stops for gas and pays with a credit card, the transaction is logged into multiple databases (O'Harrow, 1998).

15

All supported cell phones have, at a minimum, location assisting technology, and newer cell phones have assisted-GPS. The assisted-GPS cannot be turned off and will work even if the Subscriber Identity Module (SIM) card is taken out of the phone or switched with some other phone's SIM card (K.C. Jones, 2007a). (The SIM card contains the numeric codes which link the phone to the subscriber.)

The SIM card also contains a subset of the phone book data and text messages, including deleted text messages. This information can be recovered forensically, so the system can even know exactly where the user was when having sent any given message ("Sim card," n.d.). As Winston is driving along, because of the two-way communication that defines a cell phone, his location is logged by the cell phone company, possibly as often as every couple of seconds if the parameters are set that way at the cell provider's data center. This type of location and tracking mechanism is in place ostensibly in order that if Winston becomes lost or missing, the cell phone tracking records can be used to reconstruct his movements, and then to find him ("Missing Persons," 2008).

Many new cars have onboard computers which record information. Some of this information regards the condition of the engine and various onboard systems. These onboard computers also maintain continuous records of the car's speed at any given time. There are thirty data points that the federal government is going to require of all electronic data recorder (EDR) equipped cars. The EDR will be a requirement for all new cars sold starting in the 2013 model year. As of 2008, many manufacturers have started equipping their cars with EDRs. In most cases the EDRs write over old information as new information is recorded. The overwrite time is variable, but specified so that the thirty data points are available for a time long enough that in the event of a crash, the insurance company and law enforcement can determine what the car was doing just prior to a crash.

In some cars, a vehicle status data recorder, which does not overwrite data and is always on, is included. The ostensible reason for installing this device is for manufacturers to determine if and when a driver violates the warranty, and what part (or parts) is (are) affected by the behavior which would void the warranty, such as racing (Gritzinger, 2008).

Winston may need to pass through a toll booth to get to work. Several tollway systems offer toll transponders which have the toll amounts automatically taken from an account which is charged with a credit card. Each transponder is uniquely numbered so that the correct amount can be debited to the correct account for toll payment. The time and ID number of the transponder are written to a database for billing and to settle disputes. More recently the toll records have been used, not just for criminal investigations, but also in civil cases such as divorces, to prove the car was not where the lying spouse said it was at whatever time was in question (Newmarker, 2007).

The same toll lanes which offer this convenience also have cameras. The cameras photograph the vehicles that pass through the transponder lanes, and the reason given for having the cameras is to prevent fraud and misuse, and of course send tickets to those without transponders ("E-Z Pass," 2006). The fraud could consist of such instances as someone manufacturing a counterfeit toll transponder programmed with someone else's correctly guessed or deduced transponder ID number, or something as simple as the use of a stolen transponder. (Of course if the owner notices and reports the theft, the toll passing capabilities would presumably be shut off at the central computer.) Transponders can also just be hacked, with the bad guy able to read another individual's legitimate transponder and then use the transponder code (Mills, 2008).

Winston will still have his travel recorded by a camera. Many toll booths are dual use and can take cash and still read the toll transponders. The cameras are trained on cars traveling through in those lanes ostensibly to combat fraud and abuse and for crime prevention in general. Cameras for tracking cars via license plates are used in law enforcement's patrol cars.

When Winston arrives at a garage in an urban area, the car is photographed on entering the garage, and Winston pushes the button for a ticket which contains a magnetic stripe. When Winston exits the garage, if payment is made with a credit card, his identity is associated with the garage ticket (which contained the entry time and by inference a cross-reference to the pictorial record of entry) and the car's visit to and length of stay at the garage is written to a database. The stated reason given for this type of observation involves physical security (Haas & Giovis, 2008).

If Winston parked on a city street in a downtown urban area and received a ticket for staying in the space for longer than amount of time for which he paid the parking meter, that information is entered into a database ("Parking Ticket Management Solutions," 2008). There had been a surveillance effect of police department parking patrol personnel patrolling the streets with handheld computer-like wireless devices, but that was the old way. Today, the fact is that many cities have instituted laws which require cars which have more than a certain number of outstanding tickets logged against them to be booted, i.e., fitted with the "Denver boot." Parking patrol in Chicago canvasses the streets using twenty-six vans, each equipped with LPRs and checking cars on both sides of the street at 1,000 cars per hour. The license plates of the cars are compared to a list of wanted license plates (Washburn, 2007). The most efficient mechanism involves transmitting the license plate of the parked car to a central database and having the computer make the database lookup from the centralized database. There is nothing to prevent the time and location of the query from being logged in the database, and many reasons to expect that the recording of the encounter will take place. Even if Winston's car is parked legally, if a parking patrol van drives by, the city or whatever agency is interested will know when and where his car was parked.

If Winston or Julia withdrew cash from an Automated Teller Machine (ATM) on their morning break, the time and location, and picture from the ATM camera is duly stored in a database ("Digital Recording," 2004). The ostensible purpose of this type of surveillance is crime and fraud prevention and detection. 25

At lunchtime, Winston may buy something for his wife with a credit card, maybe for her birthday, maybe a purse she admired. That credit card transaction, including a detailed list of items purchased, is recorded. There is likely a camera recording the transaction, and at any time in the future, that transaction's video can be accessed keyed on the financial transaction itself (Vlahos, 2008). There may be cameras in the lobby and/or trained on the building entrance at work. Those cameras will record his return to work. As of January, 2008, there were an estimated 30 million surveillance cameras in the United States recording the goings on in public and publicly accessible commercial spaces. These cameras were recording 4 billion hours of images per week (Vlahos, 2008).

When Winston swipes his card for access to restricted buildings and areas, that information is recorded in a database. Winston uses a computer to do his job. Many employers have installed monitoring software to ascertain what exactly Winston, and other employees like him, do on their computers. At the very least, the bandwidth providers know what he does on the internet. Some corporations and government agencies employ key loggers, which record every key the computer user strikes. So when Winston sends his wife an e-card for her birthday, the boss will know. Of course many employees avoid doing any personal business on the computers at work for the reason that they are monitored. The reason given by employers for using key loggers and other monitoring software is to measure employee productivity and monitor activity, perhaps to prevent the theft of insider secrets or other nefarious actions on the part of the employee ("Internet and Computer," n.d.). This type of surveillance, however, is expected, as employers merely state in the employee handbook or wherever their policies are published that the employee has no reasonable expectation of privacy when using work-related resources (Eureste, 2008).

> "It is well within the capabilities of the NSA to digitize and record every phone conversation made by every citizen."

Every phone call is logged to the telecom's databases, and recently, legal authority to conduct real-time wiretapping against wide swaths of the citizenry has been granted by the United States government to itself (Frederickson, 2008). This authority is in addition to the warrantless and illegal wiretapping which occurred starting in 2003 and in some limited form even prior to that, in 2001 and before (Bamford, 2008).

The scope of the illegal surveillance and the number of calls illegally wiretapped will never be known. The FISA Amendment Act gave the telecoms retroactive immunity for illegal actions committed by the telecoms which were in violation of the FISA law as it had stood at the time the crimes were committed. Mark Klein, the AT&T employee who came forward with evidence of the criminality offered to testify to Congress about the crimes he witnessed. He was never called to testify by any committee of Congress. His lawyer's letters offering the testimony were never answered. He eventually said, "There will never be any hearings. It will die, and you'll never find out what they did" (Goodman & Klein, 2008).

The supposed reason for wiretapping Americans and storing the phone numbers citizens dialed, citizens who were under no reasonable suspicion and for which the government certainly had no probable cause, was that this type of Fourth Amendment violation was necessary for anti-terrorism purposes. So Winston Smith calls his accountant, and the phone number he calls is entered into the database. Quite possibly his conversation was recorded. It is well within the capabilities of the NSA to digitize and record every phone conversation made by every citizen.

References

Bamford, J. (2008). *The Shadow Factory.* New York: Doubleday.

Charny, B. (2004, September 29). Janet Jackson still holds TiVo title. Retrieved January 7, 2009, from http://news.cnet.com/Janet-Jackson-still-holds-TiVo-title/2100-1041_3-5388626.html

Data Mining with MicroStrategy: Using the MicroStrategy BI Platform to Distribute Data Mining to the Masses. (n.d.). *Bitpipe* Retrieved October 16, 2010

Deleting Web Browser Cookies & Cache. (2006, August 31). *New York University Information Technology Services* Retrieved January 7, 2009, from http://www.nyu.edu/its/faq/cache.html

Digital recording with Dibos in bank applications. (2004, March 3). *Bosch* Retrieved January 9, 2009, from http://resource.boschsecurity.com/documents/DiBos19InchDigi_ApplicationReference_Bank_enUS_T2822415627.pdf

E-Z Pass Toll System. (2006, January 18). *import rival* Retrieved January 7, 2009, from http://www.importrival.com/modules/AMS/article.php?storyid=37

Eureste, M.A. (2008, December 18). Steps to Avoid Violating Employee Privacy Rights. *HR Tools* Retrieved January 9, 2009, from http://www.hrtools.com/insights/mary_alice_eureste/steps_to_avoid_violating_employee_privacy_rights.aspx

Fox, B. (2006, April 26). Invention: Apple's all-seeing screen. Retrieved November 7, 2008, from http://www.newscientist.com/article.ns?id=dn9059&print=true

Frederickson, S. (2008). Tapping into the reporter's notebook. *The News Media & the Law, 32*(4).

Godfrey, S. (2008, December 3). Nobody Rides for Free. *Washington City Paper* Retrieved February 22, 2009, from http://www.washingtoncitypaper.com/display.php?id=36563

Goodman, A., & Klein, M. (2008, July 7). AT&T Whistleblower Urges Against Immunity for Telecoms in Bush Spy Program. *Democracy Now* Retrieved January 9, 2009, from http://i4.democracynow.org/2008/7/7/att_t_whistleblower_urges_against_immunity

Gritzinger, B. (2008, September 26). Black box on board. *Auto Week* Retrieved January 8, 2009, from http://www.autoweek.com/apps/pbcs.dll/article?AID=/20080924/FREE/809189970/1023/THISWEEKSISSUE

Haas, B., & Giovis, J. (2008, May 20). Crime can strike fast at South Florida malls. *TCPalm* Retrieved February 15, 2009, from http://www.tcpalm.com/news/2008/may/20/crime-can-strike-fast-south-florida-malls/

Intelligent Transportation Systems (ITS). (2005, March 18). *United We Ride* Retrieved January 7, 2009, from http://www.unitedweride.gov/MMS-ITS-3-18-05.doc

Internet and Computer Monitoring Software. (n.d.). *Workexaminer* Retrieved January 9, 2009, from http://www.workexaminer.com/

Jones, K. C. (2007a, January 22). Thieves Busted By GPS-Enabled Booty *Information Week* Retrieved February 25, 2009, from http://www.informationweek.com/story/showArticle.jhtml?articleID=196902643

Million Dollar Netflix. (2006, October 2). *Kiosk.net* Retrieved January 7, 2009, from http://kiosk.net/2006/10/million-dollar-netflix/

Mills, E. (2008, August 6). Hacking electronic-toll systems. *CNET News* Retrieved January 7, 2009, from http://news.cnet.com/8301-1009_3-10009353-83.html

Missing Persons Investigative Best Practices Protocol Unidentified Deceased Persons Investigative Guidelines. (2008, October 30). *New Jersey State Police* Retrieved January 9, 2009, from http://www.njsp.org/divorg/invest/pdf/mpi-best-practices-protocol-103008.pdf

Nakashima, E. (2008a, January 26). Bush Order Expands Network Monitoring. *WashingtonPost.com* Retrieved January 7, 2009, from http://www.washingtonpost.com/wp-dyn/content/article/2008/01/25/AR2008012503261.html?wpisrc=rss_technology

Newmarker, C. (2007, August 10). E-ZPass records out cheater in divorce court. *msnbc* Retrieved January 6, 2009, from http://www.msnbc.msn.com/id/20216302/

O'Harrow, R. (1998, March 8). Are Data Firms Getting Too Personal. *Washington Post.* Retrieved November 6, 2008, from http://www.washingtonpost.com/wp-srv/frompost/march98/privacy8.htm

Parking Ticket Management Solutions. (2008, August 19). *Complus Data Innovations, Inc.* Retrieved January 6, 2009, from http://www.complusdata.com/news.asp

Sim card data recovery Software. (n.d.). *datadoctor.org* Retrieved January 8, 2009, from http://www.datadoctor.org/partition-recovery/sim-card.html

Singel, R. (2007, June 12). AT&T 'Spy Room' Documents Released, Confirm Wired News' Earlier Publication. *Wired* Retrieved February 22, 2009, from http://blog.wired.com/27bstroke6/2007/06/att_spy_room_do.html

SmarTrip Questions & Answers. (n.d.). *WMATA* Retrieved October 17, 2010, from http://www.wmata.com/fares/smartrip/smartrip_qanda.cfm#balance_protected

Tortora, V. R. (1998). The Seventh Sense. *Focus, 45*(1), 15–18.

Understanding the Text

1. According to Eyre, which types of technologies are storing the most information on the average citizen? Are these technologies most citizens use?
2. How many moments did Winston allow his data to be monitored? Was there anything he could have avoided? If so, how?

Reflection and Response

3. Eyre explains the various ways an individual's data is gathered and stored. Think about the technologies you use every day. What types of data do you think may have been collected? Does it bother you? Why or why not?
4. Eyre says that if someone visits a government website, the government is legally allowed to record any information that "traverses a federal government network" (par. 8). The government says this is to prevent cyberattacks on the nation. That makes sense, but are there potential problems with this? Could this violate any laws or rights? Explain.

Making Connections

5. "As of January, 2008, there were an estimated 30 million surveillance cameras in the United States recording the goings on in public and publicly accessible commercial spaces," explains Eyre (par. 26). Now, more than a decade later, it's safe to assume that there are even more. Research the Fourth Amendment, surveillance rights, and other privacy laws. Do these ubiquitous surveillance cameras violate any laws or rights? Why or why not?
6. Eyre argues that "it is well within the capabilities of the NSA to digitize and record every phone conversation made by every citizen" (par. 28). Spend time researching this topic, making sure you read arguments both for and against the NSA's spying capabilities. Which side has a stronger argument? Why?

Facebook Faces a New World as Officials Rein in a Wild Web

Paul Mozur, Mark Scott, and Mike Isaac

Paul Mozur is a reporter based in Hong Kong, where he writes about technology and cybersecurity. Mark Scott has been a writer since 2005, and today, he works for Politico as their chief of technology correspondent. Mike Isaac is a tech reporter, too, whose work focuses on up-and-coming tech companies. Together, authors Mozur, Scott, and Isaac write about topics ranging from cybersecurity to high-profile companies like Facebook and Uber.

Here, the authors discuss the challenges Facebook faces as it tries to expand internationally. According to the article, over fifty countries have begun enacting laws in an effort to control how their citizens use the World Wide Web. Facebook, in particular, has had a contentious relationship with various countries, most notably China. The authors explain how Facebook is trying to work its way into every country and why some countries are doing all they can to stop it.

On a muggy, late spring evening, Tuan Pham awoke to the police storming his house in Hanoi, Vietnam. They marched him to a police station and made their demand: Hand over your Facebook password. Mr. Tuan, a computer engineer, had recently written a poem on the social network called "Mother's Lullaby," which criticized how the communist country was run.

One line read, "One century has passed, we are still poor and hungry, do you ask why?"

Mr. Tuan's arrest came just weeks after Facebook offered a major olive branch to Vietnam's government. Facebook's head of global policy management, Monika Bickert, met with a top Vietnamese official in April and pledged to remove information from the social network that violated the country's laws.

While Facebook said its policies in Vietnam have not changed, and it has a consistent process for governments to report illegal content, the Vietnamese government was specific. The social network, they have said, had agreed to help create a new communications channel with the government to prioritize Hanoi's requests and remove what the regime considered inaccurate posts about senior leaders.

5 Populous, developing countries like Vietnam are where the company is looking to add its next billion customers—and to bolster its ad business. Facebook's promise to Vietnam helped the social media giant

placate a government that had called on local companies not to advertise on foreign sites like Facebook, and it remains a major marketing channel for businesses there.

The diplomatic game that unfolded in Vietnam has become increasingly common for Facebook. The internet is Balkanizing, and the world's largest tech companies have had to dispatch envoys to, in effect, contain the damage such divisions pose to their ambitions. The internet has long had a reputation of being an anything-goes place that only a few nations have tried to tame—China in particular. But in recent years, events as varied as the Arab Spring, elections in France and confusion in Indonesia over the religion of the country's president have awakened governments to how they have lost some control over online speech, commerce and politics on their home turf. Even in the United States, tech giants are facing heightened scrutiny from the government. Facebook recently cooperated with investigators for Robert S. Mueller III, the special counsel investigating Russian interference in the American presidential election. In recent weeks, politicians on the left and the right have also spoken out about the excess power of America's largest tech companies.

As nations try to grab back power online, a clash is brewing between governments and companies. Some of the biggest companies in the world—Google, Apple, Facebook, Amazon and Alibaba among them—are finding they need to play by an entirely new set of rules on the once-anarchic internet.

And it's not just one new set of rules. According to a review by the *New York Times*, more than fifty countries have passed laws over the last five years to gain greater control over how their people use the web.

"Ultimately, it's a grand power struggle," said David Reed, an early pioneer of the internet and a former professor at the M.I.T. Media Lab. "Governments started waking up as soon as a significant part of their powers of communication of any sort started being invaded by companies."

Facebook encapsulates the reasons for the internet's fragmentation—and increasingly, its consequences.

The company has become so far-reaching that more than two billion people—about a quarter of the world's population—now use Facebook each month. Internet users (excluding China) spend one in five minutes online within the Facebook universe, according to comScore, a research firm. And Mark Zuckerberg, Facebook's chief executive, wants that dominance to grow.

But politicians have struck back. China, which blocked Facebook in 2009, has resisted Mr. Zuckerberg's efforts to get the social network back into the country. In Europe, officials have repudiated Facebook's attempts to gather data from its messaging apps and third-party websites.

The Silicon Valley giant's tussle with the fracturing internet is poised to escalate. Facebook has now reached almost everyone who already has some form of internet access, excluding China. Capturing those last users—including in Asian nations like Vietnam and African countries like Kenya—may involve more government roadblocks. "We understand that and accept that our ideals are not everyone's," said Elliot Schrage, Facebook's vice president of communications and public policy. "But when you look at the data and truly listen to the people around the world who rely on our service, it's clear that we do a much better job of bringing people together than polarizing them."

Friending China

By mid-2016, a yearslong campaign by Facebook to get into China—the world's biggest internet market—appeared to be sputtering.

Mr. Zuckerberg had wined and dined Chinese politicians, publicly showed off his newly acquired Chinese-language skills—a moment that set the internet abuzz—and talked with a potential Chinese partner about pushing the social network into the market, according to a person familiar with the talks who declined to be named because the discussions were confidential. 15

At a White House dinner in 2015, Mr. Zuckerberg had even asked the Chinese president, Xi Jinping, whether Mr. Xi might offer a Chinese name for his soon-to-be-born first child—usually a privilege reserved for older relatives, or sometimes a fortune teller. Mr. Xi declined, according to a person briefed on the matter.

But all those efforts flopped, foiling Facebook's attempts to crack one of the most isolated pockets of the internet.

China has blocked Facebook and Twitter since mid-2009, after an outbreak of ethnic rioting in the western part of the country. In recent years, similar barriers have gone up for Google services and other apps, like Line and Instagram.

Even if Facebook found a way to enter China now, it would not guarantee financial success. Today, the overwhelming majority of Chinese citizens use local online services like Qihoo 360 and Sina Weibo. No American-made apps rank among China's fifty most popular services, according to SAMPi, a market research firm.

Chinese tech officials said that although many in the government are open to the idea of Facebook releasing products in China, there is resistance among leaders in the standing committee of the country's Politburo, its top decision-making body. 20

In 2016, Facebook took tentative steps toward embracing China's censorship policies. That summer, Facebook developed a tool that could suppress posts in certain geographic areas, the *Times* reported last year. The idea was that it would help the company get into China by enabling Facebook or a local partner to censor content according to Beijing's demands. The tool was not deployed.

In another push last year, Mr. Zuckerberg spent time at a conference in Beijing that is a standard on the China government relations tour. Using his characteristic brand of diplomacy—the Facebook status update—he posted a photo of himself running in Tiananmen Square on a dangerously smoggy day. The photo drew derision on Twitter, and concerns from Chinese about Mr. Zuckerberg's health.

For all the courtship, things never quite worked out.

"There's an interest on both sides of the dance, so some kind of product can be introduced," said Kai-Fu Lee, the former head of Google in China who now runs a venture-capital firm in Beijing. "But what Facebook wants is impossible, and what they can have may not be very meaningful."

This spring, Facebook tried a different tactic: testing the waters in China without telling anyone. The company authorized the release of a photo-sharing app there that does not bear its name, and experimented by linking it to a Chinese social network called WeChat. One factor driving Mr. Zuckerberg may be the brisk ad business that Facebook does from its Hong Kong offices, where the company helps Chinese companies—and the government's own propaganda organs—spread their messages. In fact, the scale of the Chinese government's use of Facebook to communicate abroad offers a notable sign of Beijing's understanding of Facebook's power to mold public opinion.

Chinese state media outlets have used ad buys to spread propaganda around key diplomatic events. Its stodgy state-run television station and the party mouthpiece newspaper each have far more Facebook "likes" than popular Western news brands like CNN and Fox News, a likely indication of big ad buys.

To attract more ad spending, Facebook set up one page to show China's state broadcaster, CCTV, how to promote on the platform, according to a person familiar with the matter. Dedicated to Mr. Xi's international trips, the page is still regularly updated by CCTV, and has 2.7 million likes. During the 2015 trip when Mr. Xi met Mr. Zuckerberg, CCTV used the channel to spread positive stories. One post was titled "Xi's UN address wins warm applause."

Fittingly, Mr. Zuckerberg's eagerness and China's reluctance can be tracked on Facebook.

In November 2018, the U.K. Parliament held a meeting with representatives from nine countries with the intention of asking Mark Zuckerberg about data harvesting on his platform that led to the rampant spread of fake news on Facebook. Zuckerberg did not attend, instead sending a vice president of the company, resulting in much criticism of Zuckerberg.
Jack Taylor/Getty Images

During Mr. Xi's 2015 trip to America, Mr. Zuckerberg posted about how the visit offered him his first chance to speak a foreign language with a world leader. The post got more than a half million likes, including from Chinese state media (despite the national ban). But on Mr. Xi's propaganda page, Mr. Zuckerberg got only one mention—in a list of the many tech executives who met the Chinese president.

Europe's Privacy Pushback

Last summer, e-mails winged back and forth between members of Facebook's global policy team. They were finalizing plans, more than two years in the making, for WhatsApp, the messaging app Facebook had bought in 2014, to start sharing data on its one billion users with its new parent company. The company planned to use the data to tailor ads on Facebook's other services and to stop spam on WhatsApp. 30

A big issue: how to win over wary regulators around the world.

Despite all that planning, Facebook was hit by a major backlash. A month after the new data-sharing deal started in August 2016, German privacy officials ordered WhatsApp to stop passing data on its 36 million local users to Facebook, claiming people did not have enough say over how it would be used. The British privacy watchdog soon followed.

By late October, all twenty-eight of Europe's national data-protection authorities jointly called on Facebook to stop the practice. Facebook quietly mothballed its plans in Europe. It has continued to collect people's information elsewhere, including the United States.

"There's a growing awareness that people's data is controlled by large American actors," said Isabelle Falque-Pierrotin, France's privacy regulator. "These actors now know that times have changed."

Facebook's retreat shows how Europe is effectively employing regulations—including tough privacy rules—to control how parts of the internet are run.

The goal of European regulators, officials said, is to give users greater control over the data from social media posts, online searches and purchases that Facebook and other tech giants rely on to monitor our online habits.

"The goal of European regulators, officials said, is to give users greater control over the data from social media posts, online searches and purchases that Facebook and other tech giants rely on to monitor our online habits."

As a tech company whose ad business requires harvesting digital information, Facebook has often underestimated the deep emotions that European officials and citizens have tied into the collection of such details. That dates back to the time of the Cold War, when many Europeans were routinely monitored by secret police.

Now, regulators from Colombia to Japan are often mimicking Europe's stance on digital privacy. "It's only natural European regulators would be at the forefront," said Brad Smith, Microsoft's president and chief legal officer. "It reflects the importance they've attached to the privacy agenda."

In interviews, Facebook denied it has played fast and loose with users' online information and said it complies with national rules wherever it operates. It questioned whether Europe's position has been effective in protecting individuals' privacy at a time when the region continues to fall behind the United States and China in all things digital.

Still, the company said it respected Europe's stance on data protection, particularly in Germany, where many citizens have long memories of government surveillance.

"There's no doubt the German government is a strong voice inside the European community," said Richard Allan, Facebook's head of public policy in Europe. "We find their directness pretty helpful."

Europe has the law on its side when dictating global privacy. Facebook's non-North American users, roughly 1.8 billion people, are primarily overseen by Ireland's privacy regulator because the company's international headquarters is in Dublin, mostly for tax reasons. In 2012, Facebook was forced to alter its global privacy settings—including those in the United States—after Ireland's data protection watchdog found problems while auditing the company's operations there.

Three years later, Europe's highest court also threw out a fifteen-year-old data-sharing agreement between the region and the United States following a complaint that Facebook had not sufficiently protected Europeans' data when it was transferred across the Atlantic. The company denies any wrongdoing.

And on September 12, Spain's privacy agency fined the company 1.2 million euros for not giving people sufficient control over their data when Facebook collected it from third-party websites. Watchdogs in Germany, the Netherlands and elsewhere are conducting similar investigations. Facebook is appealing the Spanish ruling.

"Facebook simply can't stick to a one-size-fits-all product around the world," said Max Schrems, an Austrian lawyer who has been a Facebook critic after filing the case that eventually overturned the fifteen-year-old data deal. 45

Potentially more worrying for Facebook is how Europe's view of privacy is being exported. Countries from Brazil to Malaysia, which are crucial to Facebook's growth, have incorporated many of Europe's tough privacy rules into their legislation.

"We regard the European directives as best practice," said Pansy Tlakula, chairwoman of South Africa's Information Regulator, the country's data protection agency. South Africa has gone so far as to copy whole sections, almost word-for-word, from Europe's rule book.

The Play for Kenya

Blocked in China and troubled by regulators in Europe, Facebook is trying to become "the internet" in Africa. Helping get people online, subsidizing access, and trying to launch satellites to beam the internet down to the markets it covets, Facebook has become a dominant force on a continent rapidly getting online.

But that has given it a power that has made some in Africa uncomfortable.

Some countries have blocked access, and outsiders have complained Facebook could squelch rival online business initiatives. Its competition with other internet companies from the United States and China has drawn comparisons to a bygone era of colonialism. For Kenyans 50 like Phyl Cherop, 33, an entrepreneur in Nairobi, online life is already dominated by the social network. She abandoned her bricks-and-mortar store in a middleclass part of the city in 2015 to sell on Facebook and WhatsApp.

"I gave it up because people just didn't come anymore," said Ms. Cherop, who sells items like designer dresses and school textbooks. She added that a stand-alone website would not have the same reach. "I prefer using Facebook because that's where my customers are. The first thing people want to do when they buy a smartphone is to open a Facebook account."

As Facebook hunts for more users, the company's aspirations have shifted to emerging economies where people like Ms. Cherop live. Less than 50 percent of Africa's population has internet connectivity, and regulation is often rudimentary.

Since Facebook entered Africa about a decade ago, it has become the region's dominant tech platform. Some 170 million people—more than two thirds of all internet users from South Africa to Senegal—use it, according Facebook's statistics. That is up 40 percent since 2015.

The company has struck partnerships with local carriers to offer basic internet services—centered on those offered by Facebook—for free. It has built a pared-down version of its social network to run on the cheaper, less powerful phones that are prevalent there.

Facebook is also investing tens of millions of dollars alongside telecom operators to build a 500-mile fiber-optic internet connection in rural Uganda. In total, it is working with about thirty regional governments on digital projects.

"We want to bring connectivity to the world," said Jay Parikh, a Facebook vice president for engineering who oversees the company's plans to use drones, satellites and other technology to connect the developing world.

Facebook is racing to gain the advantage in Africa over rivals like Google and Chinese players including Tencent, in a twenty-first-century version of the "Scramble for Africa." Google has built fiber internet networks in Uganda and Ghana. Tencent has released WeChat, its popular messaging and e-commerce app, in South Africa.

Facebook has already hit some bumps in its African push. Chad blocked access to Facebook and other sites during elections or political protests. Uganda also took legal action in Irish courts to force the social network to name an anonymous blogger who had been critical of the government. Those efforts failed.

In Kenya, one of Africa's most connected countries, there has been less pushback.

Facebook expanded its efforts in the country of 48 million in 2014. It teamed up with Airtel Africa, a mobile operator, to roll out Facebook's Free Basics—a no-fee version of the social network, with access to certain news, health, job and other services there and in more than twenty other countries worldwide. In Kenya, the average person has a budget of just 30 cents a day to spend on internet access.

Free Basics now lets Kenyans use Facebook and its Messenger service at no cost, as well as read news from a Kenyan newspaper and view information about public health programs. Joe Mucheru, Kenya's tech minister, said it at least gives his countrymen a degree of internet access.

Still, Facebook's plans have not always worked out. Many Kenyans with access to Free Basics rely on it only as a backup when their existing smartphone credit runs out.

"Free Basics? I don't really use it that often," said Victor Odinga, 27, an accountant in downtown Nairobi. "No one wants to be seen as someone who can't afford to get online."

Understanding the Text

1. What did Facebook do in 2016 in an attempt to get China to make access available to its citizens? Why do you think the tool was never deployed?
2. Why is Europe, generally speaking, apprehensive about Facebook expanding its presence on the continent? Is Europe's apprehension justified or over the top? Explain.

Reflection and Response

3. The authors say that "Europe is effectively employing regulations—including tough privacy rules—to control how parts of the internet are run" (par. 35). What are some potential benefits and disadvantages to a government taking this action? Explain.
4. The article explains how one Kenyan woman is using Facebook and WhatsApp to run an online business. In fact, an estimated 170 million people currently use Facebook in Africa (par. 53). That said, the authors discuss how the network has been blocked or used as by the government as an investigative tool during political unrest in various African countries. According to this article's evidence, does Facebook do more harm than good? Explain your stance using textual evidence.

Making Connections

5. On his trip to China, Mark Zuckerberg posted a picture of himself running through Tiananmen Square. Research what happened in Tiananmen Square. Why might Zuckerberg, knowing the history of the location, post that photo? Does this explain China's resistance to Facebook and a truly free internet in China? Explain.
6. Facebook, generally speaking, has not faced the type of backlash or pushback in the United States as it has in other countries. Research issues or challenges Facebook has faced around the world. Then, write an argument-driven essay explaining why you believe, using credible research, Facebook has met less resistance in America compared to other countries.

Why All This Fuss about Privacy?; from *You: For Sale: Protecting Your Personal Data and Privacy Online*

Stuart Sumner

Stuart Sumner is a journalist, presenter, and editorial director at the U.K.-based magazine *Computing*, a publication for IT professionals. His expertise resides in technology, social media, and security. Sumner holds a bachelor's degree in law from the University of Southampton, and he appears regularly on the BBC as a technology pundit.

His book, *You: For Sale* (2015), details the myriad of ways everyday citizens' information is tracked, recorded, analyzed, and often sold. Sometimes, he illustrates, we share our information unintentionally by simply driving by a CCTV camera. Other times, however, we willingly give away our personal information by merely engaging in social media. The following excerpts from his book's first chapter investigate where our privacy fears began and the ways people unassumingly share their personal information with the world on a daily basis.

Why All This Fuss about Privacy?

Does privacy really matter? In fact, do we all agree on what it actually is? As Paul Sieghart said in his 1975 book *Privacy and Computers*, privacy is neither simple nor well defined. "A full analysis of all its implications needs the skills of the psychologist, the anthropologist, the sociologist, the lawyer, the political scientist and ultimately the philosopher," wrote Sieghart.

For our purposes we'll dispense with the committee necessary for this full analysis, and start with a summary of the definition which came out of the International Commission of Jurists in Stockholm in May 1967:

"The right to privacy is the right to be let alone to live one's own life with the minimum degree of interference."

That statement may be around fifty years old, but it still hits the right note. Then there's the Universal Declaration of Human Rights, signed by the UN a few years earlier in 1948:

"No one shall be subjected to arbitrary interference with his privacy, family, home, or correspondence." 5

Privacy is also enshrined in national laws. Under U.S. Constitutional law it's considered to be the right to make personal decisions regarding intimate matters (defined as issues around faith, moral values, political affiliation, marriage, procreation, or death).

Under U.S. Common Law (which is the law to come out of the courtroom—where legal precedents are routinely set and followed), privacy is defined as the right of people to lead their lives in a manner that is reasonably secluded from public scrutiny, whether that scrutiny comes from a neighbor, an investigator, or a news photographer for instance.

Finally, under U.S. statutory law, privacy is the right to be free from unwarranted drug testing and electronic surveillance.

In the U.K. the legal protection of privacy comes from the "Privacy and the Human Rights Act 1998," which basically corresponds to the rights conferred under the 1948 UN declaration.

10 As we shall see, literally every aspect of these definitions of privacy is under attack from both public and private organizations today. Individuals' privacy is being eroded at an alarming rate. In fact, some commentators are even beginning to ask if there can be such a thing as privacy in a world of cookies, government surveillance, big data and smart cities (to name just a handful of recent technology trends in part culpable for the situation).

And it's not just commentators and experts who are fretting about the privacy-free world we appear to be unwittingly signing up to. The results of the Pew Research Internet Project of 2014 reveal that 91 percent of Americans believe that consumers have lost control of their personal information. They trust the government only slightly more; 80 percent agreed that Americans should be concerned about the government's monitoring of their communications.

A White House Big Data Survey from May 2014 shows that 55 percent of respondents in the EU and 53 percent in the U.S. see the collection of big data (which can be used to identify individuals from supposedly anonymized data) as a negative.

Privacy is an aspect of our freedom. It's about being free to think and act with autonomy, and where desired, without those thoughts and actions being broadcast to others.

This book aims to explain how and why privacy is under threat, and give some basic recommendations for individuals, corporations and governments to follow in order to arrest this slide towards a world that is less free.

> "Rovio, the developers behind Angry Birds, admits that the data it collects may include but is not limited to your e-mail, device ID, IP address and location."

Here's My Cow, Now Where's My Change?

What's better than a great product at a reasonable price? What about a great product for free? It sounds too good to be true, and it is, and yet that's the lie that's repeatedly sold to all of us who use the internet or download apps to our tablets and smartphones. It's such an obvious duplicity that you'd think more of us would see through it. That we don't is a result of the way our brains are wired, but before we come to that, let's have a brief look at the path humanity has taken in its history to bring us to this point.

Next time you pass a dairy farm, you could surprise and delight your friends by remarking that it's an early form of bank (neither surprise nor delight guaranteed). That's because the earliest form of currency is thought to be cattle. If you don't possess the ability to mint coins or at least make something that's easy and cheap to reproduce accurately yet hard to copy, a cow is a pretty good substitute. As something which produces milk and can be slaughtered for meat, leather and various other products, it has its own value. Despite small variances in size, one cow is pretty much as valuable as another. You can even get change from a cow—in many early economic systems it was considered to be worth two goats.

This worked well enough for many societies from about 9,000 BC to 1,200 BC, until they were replaced by Cowrie shells. After all, it's great having a form of currency you can eat when times get tough, but it's less good when your entire life savings gets wiped out by disease or is devoured by a pack of wolves or even hungry neighbors.

Cowrie shells—egg-shaped shells belonging to a species of sea snail common to the coastal waters of the Indian and Pacific Oceans—were popular because they were almost impossible to forge, portable, and neither too rare to stymie trading, nor so common that even an egg was worth several wheelbarrow-loads (and if that was the exchange rate, how would anyone have transported enough shells to afford the wheelbarrow?). The classical Chinese character for money (貝) originated as a stylized drawing of a cowrie shell, and they are still used today in Nepal in a popular gambling game.

Things became more recognizable to us in modern societies in 1,000 BC when China first began to manufacture bronze and copper coins, originally designed to resemble Cowrie shells to help people grasp what they were for. Five hundred years later came electrum (an alloy of gold and silver) coins in Sardis, the capital city of ancient Lydia (an area which roughly corresponds to Turkey today).

Paper money first appeared in China in 806 AD, during the Tang dynasty. It's interesting to wonder what those financial pioneers would have made of the fact that their idea endures over 1,400 years later.

You might think your credit card is a relatively recent invention, but plastic money has been used in various forms since the late nineteenth century when celluloid "charge coins" were used by some hotels and department stores to enable transactions to be charged to their clients' accounts.

Credit cards as we know them now however were first used in September 1958 when Bank of America launched the somewhat unimaginatively named "BankAmericard" in Fresno, California. This eventually became the first successful recognizably modern credit card and nineteen years later changed its name to Visa.

That's the past, but what of the future of money? Some believe it to be digital currency, the most recognizable example of which today is Bitcoin, an open-source, online software payment system unaffiliated to any central authority or bank. Anyone can download the software and use their computer to help verify and record payments into a public ledger, sometimes being rewarded with bitcoins themselves for their efforts.

Others see potential in the smartphone as a wallet, with its near-field communication capabilities enabling transactions at the push of a button or simple wave in the air. If you've ever suffered the inconvenience of losing your smartphone, or worse having it stolen, then the prospect if it also being your wallet might leave you with a knotted feeling in your stomach and an empty feeling in your bank account, in which case you might prefer the idea of a chip embedded under your skin, or even a barcode branded onto your arm.

25

There are apps which claim to offer simpler ways of transacting money, and many mainstream services are now diversifying into cash transfers: Facebook, Twitter and Snapchat to name a few.

These would be free services, because nothing is precisely the amount modern online consumers expect to pay for anything. So how would these offerings make money? One way is serving advertising, and another is to sell the data they collect on their customers. Information like what size of transaction you tend to make, when you tend to make them, and what you tend to buy. Snapchat in particular has a sketchy recent history as a guardian of personal data. The popular app (and "popular" hardly does it justice, with over 700 million photos and videos shared each day) allows users to send data in the form of pictures, videos and written messages to one another. So far so pedestrian, but the supposed unique selling point is that the pictures are permanently deleted after a few seconds. Neither the receiver of the image, nor Snapchat itself can retrieve the picture once it has been deleted. Or at least, that's the idea. In practice it's the work of a moment for a savvy user to make a permanent record of whatever appears on their screen—for example by activating the phone's screenshot mode, or by using a third party app specifically designed for the task (an example of which is SnapBox, an app designed to allow users to keep Snapchat images without the sender's knowledge). And worse, in October 2014 a hacker published 12.6 gigabytes of images stolen from Snapchat's users, none of which were ever supposed to have been recorded in the first place.

However in this instance Snapchat itself isn't to blame. The breach itself occurred when now-defunct third-party site Snapsaved.com—which contained archived photos and videos from some Snapchat users—was attacked. However, Snapchat itself is guilty of a failure to act when in August 2013 Australian security firm Gibson Security alerted it to a vulnerability. In December that same year Snapchat finally put what it called mitigating features in place to plug the hole, but a few days later a hacking group bypassed the security measures calling them "minor obstacles," and then released 4.6 million Snapchat usernames and passwords in via a website called SnapchatDB.info. Snapchat apologized a week later.

But whichever economic system wins out, and the likelihood is that it will be some combination of all of the above, the one certainty seems to be that the illusion of a free lunch will continue.

Hey I Thought This Lunch Was Free!

Have you ever downloaded Angry Birds to a mobile device? How carefully did you read the user agreement that flashed up before you started catapulting creatures across the screen? If you didn't so much as glance at it, then you're with the vast majority of people, but in this case there is scant safety in numbers, the entire herd is being preyed upon, and most are totally oblivious.

We're picking on Angry Birds here as it's a common example, but most of this applies to many of the most common and well-known apps and other online services popular today, and no doubt to tomorrow's apps currently in development.

By the beginning of 2014 Angry Birds had been downloaded over 1.7 billion times. That's an incredible figure. According to the population clock at www.worldofmeters.info, there are over 7.2 billion people in the world at the time of writing. So that's almost a quarter of the entire population of the planet, potentially playing the game. Of course the real figure attacking green pigs with multi-colored birds is actually less than that, as many people will download the app several times to different devices, but the essential point is that an incredible number of people have downloaded that particular game, and the proportion who read the user agreement is probably significantly fewer than one in a million.

What you'd know if you read it, is that the app makes its money by taking your personal information and selling it on to advertisers. And given that your smartphone is so, well, smart, there's a real wealth of information for Angry Birds and apps like it to mine.

Your phone knows your location. It knows your routine, where you live, which coffee shop you stop at on the way to work, and of course where you work. It knows the route you take on your commute. It's got the contact details for just about everyone you know, and if you use it for e-mail, it's got everything you write to them, and everything they send back to you. It contains your photos, videos, and browsing habits.

Rovio, the developers behind Angry Birds, admits that the data it collects may include but is not limited to your e-mail, device ID, IP address and location. It then sells this data on to advertisers who hope to be better able target you, and sometimes where those third parties have a presence within the app itself, perhaps via an in-app advert, then they can siphon data directly from your phone. Many apps behave the same way, most gather up way more data than they need, and the vast majority employ very poor or non-existent security. In fact, it's such a treasure trove of personal information that the security agencies have got in on the act. One of the secrets uncovered by the revelations of former CIA contractor Edward Snowden is the fact that the National Security Agency (NSA) in the U.S. and the Government Communications Headquarters (GCHQ) in the U.K. have developed capabilities to take advantage of leaky apps like Angry Birds to help them compile their dossiers on their citizens (and those of other countries).

The chances are that you use more than one app on your phone, and between them, the combination of apps and websites we all use gather just about everything we do.

Google is one of the worst offenders. Via its free webmail service "Gmail," Google happily scours what you thought were your private e-mails looking for keywords, again in order to target ads at you. And it doesn't stop there, but makes full use of the potential of its wide array of products and services. Google is also able to track you across multiple devices. For example if you use Google Maps on your smartphone whilst out and about, that data will be stored and used to help target you with ads the next time you log into Google at home.

In 2013 a group of users had had enough of Google's data gathering activities, banded together and sued the company. Their argument was that Google combs its customers' e-mails in order to extract a broad meaning—or "thought data"—from them.

"Google creates and uses this 'thought data' and attaches it to the messages so Google can better exploit the communication's 'meaning' for commercial gain," they said in response to legal counter-action from Google designed to dismiss their case from the courts.

"Google collects and stores the 'thought data' separately from the e-mail message and uses the 'thought data' to: (1) spy on its users (and others); and, (2) amass vast amounts of 'thought data' on millions of people (secret user profiles)."

Google argued that federal and state wiretap laws exempt e-mail providers from liability, as it's a basic tenet of their business. So there was no attempt to deny the claims, just a shrug of the shoulders and a "Yeah, so what?"

"These protections reflect the reality that [electronic communication service] providers like Google must scan the e-mails sent to and from their systems as part of providing their services," Google said in its motion.

But the plaintiffs added a further point, and a crucial one. Google does not disclose its "thought data" mining activities to anyone. It's one thing to take someone's data in exchange for a valuable service, it's quite another to do it without permission.

"Google's undisclosed processes run contrary to its expressed agreements. Google even intercepts and appropriates the content of minors' e-mails despite the minors' legal incapacity to consent to such interception and use. Thus, these undisclosed practices are not within the ordinary course of business and cannot form the basis of informed consent," the plaintiffs said.

In March 2014, the plaintiffs lost their case. Lucy Koh, a federal judge in California, ruled in favor of Google, at the same time handing a potent defense to every company which takes its users data without their express consent.

There was a similar case in the U.K. in 2013. A group of internet users sued Google through law firm Olswang, complaining that the search giant had installed cookies (small files designed to track when someone visits a website, and what they do there) on their desktops and mobile devices despite their expressed preference to avoid precisely that activity. The individuals had used a feature in Apple's Safari browser to block third-party cookies.

In this case Google had gone way beyond simply obfuscating its intentions—it had effectively hacked its own users! For a firm whose motto is "don't be evil," it doesn't appear to be trying very hard to be good.

Google uses the data contained in these cookies about users' browsing habits to enable its partners to buy ads targeted at well-defined sections of society, for instance "high-earners," "gadget buyers," or "home owners." It also sells some of this data on directly to advertisers, once it has been anonymized. However, in the age of big data it's fairly trivial to identify individuals even from supposedly anonymized information.

There are legitimate reasons why you might not want advertisers to know what you've been using the internet for. A teenage girl might search for contraceptive advice, then be embarrassed when related ads come up when her parents are helping her with her homework. A more serious impact to the girl (or boy for that matter) than embarrassment could be a disinclination to search for information on contraception for precisely this reason. Or you might search for a wedding ring, or some other surprise gift for your partner. Do you want similar products then appearing when they sit down to do their own browsing? The situation wouldn't be so bad if you could opt out, but when the firm in question doesn't even deign to tell you it's happening, then there's a problem.

The Olswang case was brought by twelve Apple users, all of whom had been using Apple's Safari browser. Google was fined $22.5 million by the Federal Trade Commission (FTC) in the U.S. in late 2012 for exactly the same issue—putting cookies onto Safari users' devices—when a case was brought about by a different group.

Nick Pickles, director of civil liberties campaign group Big Brother Watch at the time, told U.K.-based newspaper *The Telegraph*: "This episode was no accident. Google tracked people when they had explicitly said they did not want to be tracked, so it's no surprise to see consumers who believe their privacy had been steamrollered by corporate greed seeking redress through the courts. 50

"This case could set a hugely important legal precedent and help consumers defend their privacy against profit-led decisions to ignore people's rights."

In August 2013 the *Independent*, another U.K.-based newspaper, reported that Google described the case as "not serious. . . . [T]he browsing habits of internet users are not protected as personal information, even when they potentially concern their physical health or sexuality."

Google had refused to acknowledge the case in the U.K., saying it would only recognize it in the U.S.

The *Independent* went on to quote Judith Vidal-Hall, a privacy campaigner and one of the claimants, who said: "Google's position on the law is the same as its position on tax: they will only play or pay on their home turf. What are they suggesting; that they will force Apple users whose privacy was violated to pay to travel to California to take action when they offer a service in this country on a .co.uk site? This matches their attitude to consumer privacy. They don't respect it and they don't consider themselves to be answerable to our laws on it."

It also quoted another claimant named Marc Bradshaw, who argued: "It seems to us absurd to suggest that consumers can't bring a claim 55 against a company which is operating in the U.K. and is even constructing a $1 billion headquarters in London.

"If consumers can't bring a civil claim against a company in a country where it operates, the only way of ensuring it behaves is by having a robust regulator. But the U.K. regulator, the Information Commissioner's Office, has said to me that all it can do is fine Google if it breaks the law, but Google clearly doesn't think that it is bound by that law."

"Fines would be useless—even if Google agreed to pay them—because Google earns more than the maximum fine in less than two hours. With no restraint Google is free to continue to invade our privacy whether we like it or not."

Understanding the Text

1. What does Sumner argue is portrayed as free, as a "lie that's repeatedly sold," but really isn't free (par. 15)? Do you agree with him? Why or why not?
2. How do those free products make money? Should that be allowed? Why or why not?

Reflection and Response

3. According to Sumner, about half of Americans "see the collection of big data (which can be used to identify individuals from supposedly anonymized data) as a negative" (par. 12). That means, then, that about half of the nation sees this as a positive or something neutral. Why? Are there benefits from this capability? Explain.
4. Sumner explains how the vast majority of people do not even look at, let alone read, the user agreements they consent to before downloading an app. These agreements detail what the companies are allowed to do with the data they collect from their users. Does this mean no one has a right to complain about how their data is used? Why or why not?

Making Connections

5. Read "How Is Bitcoin Money?" (p. 228) in Chapter 4. Compare Bitcoin to Sumner's definition of currency. What similarities and differences do you see? Is Bitcoin any different than the Cowrie shells used long ago? Why or why not?
6. According to this reading, apps like Angry Birds are often porous or leak information, and the NSA has developed capabilities to take advantage of that information, allowing the U.S. government to create folders on individual citizens as well as citizens of other countries. Research the user agreements for a few different apps. Is this legal? Or does the government need a warrant to collect data this way? If it is legal, should it be? Why or why not?

Privacy and the Imperative of Open Government; from *Privacy in the Modern Age: The Search for Solutions*

Steven Aftergood

Steven Aftergood, a prominent figure in the fight to combat government secrecy and a former regular on C-SPAN, has spent his career working to hold the CIA accountable. In fact, in 1997 he filed a lawsuit against the CIA, leading to the declassification of the CIA's budget for the first time in decades. Aftergood's efforts have been lauded by many; his awards include the James Madison Award from the American Library Association, the Public Access to Government Information Award from the American Association of Law Libraries, and others. Today, Aftergood is the director of the Federation of American Scientists Project on Government Secrecy.

The following piece appeared in *Privacy in the Modern Age* (2015), a collection of essays discussing and troubleshooting some of the threats to privacy in today's world. Here, Aftergood details the genesis of public privacy concerns, and explores actions he believes would assuage some justified public concerns.

Until recently, personal privacy might not have ranked very high on the list of reasons to favor open government; it would certainly have been eclipsed in prominence by arguments based on principles of democratic governance and government accountability. But following the unauthorized disclosure of highly classified intelligence programs to collect telephone metadata records in bulk, privacy concerns have emerged as among the most urgent and compelling drivers of secrecy reform.

Defenders of the classified bulk collection programs initially suggested that concerns about privacy were misplaced since "only metadata," not the actual contents of the communications, were being recorded. Some even argued that the deep secrecy of the bulk collection program somehow complemented and reinforced personal privacy. "You can't have your privacy violated if you don't know your privacy is violated," contended Representative Mike Rogers, the chair of the House Intelligence Committee, at an October 2013 hearing.

But such efforts to justify the classified collection programs were quickly overwhelmed by a tide of bipartisan° public criticism, and even erstwhile defenders of the status quo were soon endorsing or offering

Bipartisan: agreement or cooperation of two political parties that usually oppose each other's policies.

"I want to spill the beans, but I'm waiting till I have access to classified or sensitive beans."

This political cartoon from the *New Yorker* comically critiques government workers' ability to "spill the beans," so to speak. The cartoon implies that government employees wait until they have access to classified or sensitive information before becoming whistle-blowers and leaking government data.

Peter C. Vey/The New Yorker Collection/The Cartoon Bank

their own proposals for change. This turn of events raises important policy issues in many domains, but it has implications for the future of open government in particular.

Secrecy Precludes Public Consent

While privacy is the nominal° subject of the controversy, the core issue raised by classified bulk collection goes beyond the actual or potential infringement on privacy that it entails. Rather, the essential problem raised by secret bulk collection of telephone metadata records is the fact that the public was denied any opportunity to grant—or to withhold—its consent to this practice.

Nominal: existing in name only.

In this way, privacy concerns lead inexorably to the imperative of openness in government. The broad policy objective is not necessarily to defend personal privacy at all costs, but to ensure that national security policies that impinge on personal privacy are subject to public debate and approval. It is this dimension of public consent that was conspicuously absent in development and execution of the bulk collection program.

The Need to Invigorate Congressional Oversight

One might have thought that Congress, in its regular exercise of checks and balances, would have represented public concerns about privacy and that it would have provided the requisite opportunity for consent. Remarkably, in this case, that did not happen.

In retrospect, it is clear that the congressional intelligence committees did not accurately gauge or reflect public attitudes toward bulk collection. For reasons that require further investigation, the congressional watchdog did not bark. To the contrary, the committees appear to have been enablers of the program. Even after the director of national intelligence publicly denied before the Senate Intelligence Committee that any sort of mass collection of records involving U.S. persons was taking place, members of the committee did nothing to correct the record, though they knew this assertion to be false.

It may be that the congressional oversight committees are simply less capable of performing a contemporaneous oversight function than might have been supposed or hoped for. (The Senate Intelligence Committee has only recently completed a report on CIA interrogation activities that took place a full decade earlier.) Even today, there is little sign that the intelligence oversight committees have been chastened by the public uproar over bulk collection, or that they have been moved to engage in any kind of critical reflection on their role in the controversy. But if Congress wished to become more sensitive to the actual range of public interests and concerns, that would go some distance to improving the quality and integrity of national intelligence policy, including privacy policy.

> ". . . the essential problem raised by secret bulk collection of telephone metadata records is the fact that the public was denied any opportunity to grant—or to withhold—its consent to this practice."

There are several initial steps toward this end that could be undertaken: The professional diversity of committee staff could be expanded to include more persons with expertise in privacy and civil liberties issues. The number of open public hearings, which have been sparse in the last several years, and the number of nongovernmental witnesses could be increased. The record of congressional intelligence oversight itself, which includes historically valuable classified materials dating back to the Church Committee of the 1970s, could be usefully declassified. Any progress in this direction depends on setting a new standard of expectations for congressional performance and responsiveness, which depends in turn on the efforts of advocacy organizations, and ultimately on the public will itself.

A New Dawn for Transparency?

U.S. intelligence agencies had an unusual realization in the aftermath of the unauthorized disclosure of the bulk telephony metadata collection program: transparency is not necessarily a problem—it can serve their interests too. Declassification, the agencies discovered, can be used to correct errors in the record, can provide relevant context for public deliberation, and can help to counteract cynicism about official activities and motivations.

10

This realization has been translated into policy, up to a point, with tangible results: more classified government records about ongoing intelligence surveillance programs have recently been declassified than ever before. The number of pages of top secret records about bulk collection programs in particular that have been officially declassified is roughly double the number of pages leaked by Edward Snowden that have been published in the news media.

Several new government websites have been established to publicize and disseminate declassified intelligence records, including records pertaining to the privacy interests of U.S. persons. For the first time ever, a presidential directive on signals intelligence was issued by President Obama in unclassified form. New policy debates have taken shape on previously remote topics such as the privacy rights due to foreigners abroad against clandestine surveillance, and the suspected role of U.S. intelligence agencies in weakening public encryption standards and stockpiling known vulnerabilities.

Significantly, the director of national intelligence, James R. Clapper, has affirmed the view that it would have been prudent and proper to seek public consent for the bulk collection of call records from the start. "Had we been transparent about this from the outset right after 9/11 . . . we wouldn't have had the problem we had," DNI Clapper told the *Daily Beast*. This belated official acknowledgment that greater transparency would have benefited the intelligence community all along creates a foundation for a new conversation about what is wrong with current classification practices and what can be done to rectify them.

Institutionalizing Openness through External Review

Everyone, up to and including the president of the United States, recognizes that overclassification of information is a fact and a problem. While the problem can be diagnosed in different ways, most people will agree that the present system is biased in favor of classification. So the practical question is, what can be done to promote a more limited and more discriminating use of classification authority that yields less secrecy and that is more respectful of privacy?

One could imagine a revised executive order on national security classification that included a new prohibition against classifying in order to conceal infringements on privacy. But such formal limitations, like those prohibiting classifying to conceal violations of law or to prevent embarrassment, have not been notably effective. There is, however, another approach that may hold the key to significant, voluntary reductions in official secrecy.

15

That is to extend declassification authority beyond the circle of the original classifiers, and to subject agency classification decisions to external review and critique. In interesting ways, this is already being done. Between 1996 and 2012, an executive-branch body called the Interagency Security Classification Appeals Panel (ISCAP) completely overturned the classification judgments of executive-branch agencies in 27 percent of the cases that it reviewed, and it partially overturned classification decisions in another 41 percent of such cases.

This surprisingly productive record can be explained by considering the fact that the ISCAP, while fully committed to protecting legitimate national security interests, does not share all of the specific bureaucratic interests of the individual agencies whose classification judgments it rejected. Subjecting those individual agency classification decisions to an external evaluation (albeit still within the executive branch) has consistently yielded a reduction in secrecy. Having been validated in practice year after year, this basic principle could now be applied more systematically or to address particularized concerns.

So, for example, all U.S. intelligence classification guides—which are documents that present specific, itemized guidance on exactly what types of information are to be classified and at what level—could be delivered for independent review and critique to the Public Interest Declassification Board. The PIDB, created by statute, exists to, among other things, promote "the fullest possible public access to a thorough, accurate, and reliable documentary record of significant U.S. national security decisions and activities."

Agency classification guides could be designated for a more specifically privacy-related review by the Privacy and Civil Liberties Oversight Board.

This board could be asked to identify current intelligence community classification practices that have significant implications for personal privacy, to assess their validity, and to recommend appropriate changes in secrecy policy. There are other isolated "best practices" for classification review that already exist and that could easily be incorporated throughout the intelligence community and the executive branch as a whole.

The Department of Energy has a formal regulation (10 CFR 1045.20) under which members of the public may propose declassification of information that is classified under the Atomic Energy Act. I have made use of this regulation myself. A similar provision could be envisioned by which the public could challenge the classification of privacy-related and other national security information throughout the government. While one can already request declassification review of a particular document, the proposed approach would go beyond that to challenge the classification status of an entire topical area. 20

The current executive order on national security information allows for classification challenges, but only by security-cleared employees who already have access to the information. Naturally, the key to a successful classification challenge is that it must be reviewed impartially by someone other than the original classifier. But that is entirely achievable. In fiscal year 2012, government employees filed 402 such challenges, one-third of which were granted in whole or in part.

To cite one more example of an existing best practice that could be widely replicated, the inspector general of the Environmental Protection Agency actively solicits public suggestions for audits and investigations that it could perform. This potentially places one of the most powerful investigative tools in government at the disposal of public-interest requesters (and not only of actual whistle-blowers). Although only Congress has the power to legally compel an inspector general investigation, the EPA IG's receptiveness to a well-founded public "suggestion" could reasonably be expected to become standard practice.

In the end, of course, openness by itself does not change anything. Once the door to public participation has been opened, then anyone is free to enter.

Understanding the Text

1. What action caused personal privacy to become an important topic of conversation?
2. According to Aftergood, the biggest issue wasn't the actual collection of people's data; it was the fact people weren't doing what?

3. Aftergood notes that intelligence agencies realized transparency can be good for them, too. How so? Explain.

Reflection and Response

4. The article quotes Mike Rogers, then chair of the House Intelligence Committee, as saying, "You can't have your privacy violated if you don't know your privacy is violated" (par. 2). Is Rogers's logic flawed? Why or why not?
5. Aftergood writes that "most people will agree that the present system is biased in favor of classification" (par. 14). Do you agree? Why or why not?

Making Connections

6. Aftergood briefly references Edward Snowden in his opening paragraph; Snowden was the person who leaked classified information that detailed the government's collection of telephone metadata. Go online and find the article "Edward Snowden's Real Impact" by Jeffrey Toobin, published by the *New Yorker*. Does Toobin make a good case for prohibiting the public from weighing in on how data is collected, as Aftergood suggests should be the case? Why or why not? Does his article make Snowden look like a hero or a criminal? Explain.
7. Research the Freedom of Information Act (FOIA). Is this enough to help the public hold the government accountable to the same laws everyday citizens are required to follow? Why or why not?

Governments Turn to Commercial Spyware to Intimidate Dissidents

Nicole Perlroth

Nicole Perlroth is a regular contributor to the *New York Times* on the topic of cybersecurity. Perlroth's efforts to expose China's goal to steal military and industrial trade secrets have earned her several journalism awards. Perlroth previously served as deputy editor at *Forbes*, where she covered venture capital. Currently, she is working on a cybersecurity book entitled *This Is How They Tell Me the World Ends*.

Her article here tells a story of one man, Emirati activist Ahmed Mansoor, and his experiences being hacked and surveilled, and the resulting threats on his safety, security, and well-being, showing just how far a government can and will go to surveil one of its own citizens, even if that citizen has left the country.

San Francisco—In the last five years, Ahmed Mansoor, a human rights activist in the United Arab Emirates, has been jailed and fired from his job, along with having his passport confiscated, his car stolen, his e-mail hacked, his location tracked and his bank account robbed of $140,000. He has also been beaten, twice, in the same week. Mr. Mansoor's experience has become a cautionary tale for dissidents, journalists and human rights activists. It used to be that only a handful of countries had access to sophisticated hacking and spying tools. But these days, nearly all kinds of countries, be they small, oil-rich nations like the Emirates, or poor but populous countries like Ethiopia, are buying commercial spyware or hiring and training programmers to develop their own hacking and surveillance tools.

The barriers to join the global surveillance apparatus have never been lower. Dozens of companies, ranging from NSO Group and Cellebrite in Israel to Finfisher in Germany and Hacking Team in Italy, sell digital spy tools to governments.

A number of companies in the United States are training foreign law enforcement and intelligence officials to code their own surveillance tools. In many cases these tools are able to circumvent security measures like encryption. Some countries are using them to watch dissidents. Others are using them to aggressively silence and punish their critics, inside and outside their borders.

"There's no substantial regulation," said Bill Marczak, a senior fellow at the Citizen Lab at the University of Toronto's Munk School of Global

Affairs, who has been tracking the spread of spyware around the globe. "Any government who wants spyware can buy it outright or hire someone to develop it for you. And when we see the poorest countries deploying spyware, it's clear money is no longer a barrier."

Mr. Marczak examined Mr. Mansoor's e-mails and found that, before his arrest, he had been targeted by spyware sold by Finfisher and Hacking Team, which sell surveillance tools to governments for comparably cheap six- and seven-figure sums. Both companies sell tools that turn computers and phones into listening devices that can monitor a target's messages, calls and whereabouts.

In 2011, in the midst of the Arab Spring, Mr. Mansoor was arrested with four others on charges of insulting Emirate rulers. He and the others had called for universal suffrage. They were quickly released and pardoned following international pressure. But Mr. Mansoor's real troubles began shortly after his release. He was beaten and robbed of his car, and $140,000 was stolen from his bank account. He did not learn that he was being monitored until a year later, when Mr. Marczak found the spyware on his devices.

"It was as bad as someone encroaching in your living room, a total invasion of privacy, and you begin to learn that maybe you shouldn't trust anyone anymore," Mr. Mansoor recalled.

> "But these days, nearly all kinds of countries, be they small, oil-rich nations like the Emirates, or poor but populous countries like Ethiopia, are buying commercial spyware or hiring and training programmers to develop their own hacking and surveillance tools."

Mr. Marczak was able to trace the spyware back to the Royal Group, a conglomerate run by a member of the Al Nahyan family, one of the six ruling families of the Emirates. Representatives from the Emirates Embassy in Washington said they were still investigating the matter and did not return requests for further comment. Invoices from Hacking Team showed that through 2015, the Emirates were Hacking Team's second-biggest customers, behind only Morocco, and they paid Hacking Team more than $634,500 to deploy spyware on 1,100 people. The invoices came to light last year after Hacking Team itself was hacked and thousands of internal e-mails and contracts were leaked online.

Eric Rabe, a spokesman for Hacking Team, said his company no longer had contracts with the Emirates. But that is in large part because Hacking Team's global license was revoked this year by the Italian Ministry of Economic Development.

10 For now, Hacking Team can no longer sell its tools outside Europe and its chief executive, David Vincenzetti, is under investigation for some of those deals. New evidence suggests to Mr. Marczak that the Emirates may now be developing their own custom spyware to monitor their critics at home and abroad.

"The U.A.E. has gotten much more sophisticated since we first caught them using Hacking Team software in 2012," Mr. Marczak said. "They've clearly upped their game. They're not on the level of the United States or the Russians, but they're clearly moving up the chain."

Late last year, Mr. Marczak was contacted by Rori Donaghy, a London-based journalist who writes for the Middle East Eye, an online news site, and a founder of the Emirates Center for Human Rights, an independent organization that tracks human rights abuses in the Emirates. Mr. Donaghy asked Mr. Marczak to examine suspicious e-mails he had received from a fictitious organization called the Right to Fight. The e-mails asked him to click on links about a panel on human rights.

Mr. Marczak found that the e-mails were laden with highly customized spyware, unlike the off-the-shelf varieties he has become accustomed to finding on the computers of journalists and dissidents. As Mr. Marczak examined the spyware further, he found that it was being deployed from sixty-seven different servers and that the e-mails had baited more than 400 people into clicking its links and unknowingly loading its malware onto their machines.

He also found that twenty-four Emiratis were being targeted with the same spyware on Twitter. At least three of those targeted were arrested shortly after the surveillance began; another was later convicted of insulting Emirate rulers in absentia.

15 Mr. Marczak and the Citizen Lab plan to release details of the custom Emirates spyware online on Monday. He has developed a tool he called Himaya—an Arabic word that roughly means "protection"—that will allow others to see if they are being targeted as well.

Mr. Donaghy said he was frightened by Mr. Marczak's findings, but not surprised.

"Once you dig beneath the surface, you find an autocratic° state, with power centralized among a handful of people who have increasingly used their wealth for surveillance in sophisticated ways," Mr. Donaghy said.

The Emirates have cultivated an image as progressive allies of the United States in the Middle East. Their rulers often highlight their sizable foreign aid budget and their women's rights efforts. But human rights

Autocratic: when a ruler has absolute power.

monitors say the Emirates have been aggressive in trying to neutralize their critics.

"The U.A.E. has taken some of the most dramatic steps to shut down individual human rights activists and dissenting voices," said James Lynch, the deputy director for Amnesty International's program in the Middle East and North Africa. "It is highly sensitive to its image and fully aware of who is criticizing the country from abroad."

20 Last summer, Mr. Lynch was invited to speak about labor rights at a construction conference in Dubai and was turned away at the airport. Officials did not give a reason, but he later saw that his deportation certificate listed reasons of security.

Mr. Mansoor, who still resides in the Emirates, has been outspoken about the use of spyware but is increasingly limited in what he can do. He worries that anyone he speaks to will also become a target.

And more recently, the state has started punishing the families of those who speak out, as well. In March, the Emirates revoked the passports of three siblings whose father was charged with attempting to overthrow the state.

"You'll wake up one day and find yourself labeled a terrorist," Mr. Mansoor said.

"Despite the fact you don't even know how to put a bullet inside a gun."

Understanding the Text

1. According to the article, why is spyware easy to obtain and use?
2. Perlroth explains that sometimes families of those who speak up are punished, too. Why is that?

Reflection and Response

3. The author tells us that Bill Marczak discovered that "twenty-four Emiratis were being targeted with the same spyware on Twitter" (par. 14). Does this make you think of social media differently? Why or why not?
4. Why do you think the U.A.E., despite some of its progressive leanings, works hard to shut down critics? Does America ever do something similar?

Making Connections

5. Think back to the first reading in this chapter, "Surveillance Today" (p. 141) by William Eyre. Does Perlroth's article make Eyre's more or less frightening? Explain.
6. Read the *Wikipedia* entry on Julian Assange, paying extra attention to his involvement in leaks and hacking. Does Perlroth's article make a case for the need for more people like Assange? Explain.

Harrison Bergeron

Kurt Vonnegut

One of America's most famous writers, Kurt Vonnegut is known for his novels and short stories in the science fiction genre, including his best-selling book, *Slaughterhouse-Five* (1969), which rocketed him to fame. *Slaughterhouse-Five*, which draws on his experiences as a prisoner of war, helped propel Vonnegut as a face of the anti-war movement after World War II.

The following story is often taught in the classroom, spanning the disciplines. Set in the year 2081, Vonnegut's story propels his readers to the future, a time in which additional constitutional amendments have been implemented, demanding complete equality among American citizens. Vonnegut's story shows us what might happen if technology is employed in an effort to equalize society.

The year was 2081, and everybody was finally equal. They weren't only equal before God and the law. They were equal every which way. Nobody was smarter than anybody else. Nobody was better looking than anybody else. Nobody was stronger or quicker than anybody else. All this equality was due to the 211th, 212th, and 213th Amendments to the Constitution, and to the unceasing vigilance of agents of the United States Handicapper General.

Some things about living still weren't quite right, though. April for instance, still drove people crazy by not being springtime. And it was in that clammy month that the H-G men took George and Hazel Bergeron's fourteen-year-old son, Harrison, away.

It was tragic, all right, but George and Hazel couldn't think about it very hard. Hazel had a perfectly average intelligence, which meant she couldn't think about anything except in short bursts. And George, while his intelligence was way above normal, had a little mental handicap radio in his ear. He was required by law to wear it at all times. It was tuned to a government transmitter. Every twenty seconds or so, the transmitter would send out some sharp noise to keep people like George from taking unfair advantage of their brains.

George and Hazel were watching television. There were tears on Hazel's cheeks, but she'd forgotten for the moment what they were about.

On the television screen were ballerinas. 5

A buzzer sounded in George's head. His thoughts fled in panic, like bandits from a burglar alarm.

"That was a real pretty dance, that dance they just did," said Hazel.

"Huh" said George.

"That dance—it was nice," said Hazel.

"Yup," said George. He tried to think a little about the ballerinas. They weren't really very good—no better than anybody else would have been, anyway. They were burdened with sashweights and bags of birdshot, and their faces were masked, so that no one, seeing a free and graceful gesture or a pretty face, would feel like something the cat drug in. George was toying with the vague notion that maybe dancers shouldn't be handicapped. But he didn't get very far with it before another noise in his ear radio scattered his thoughts. 10

George winced. So did two out of the eight ballerinas.

Hazel saw him wince. Having no mental handicap herself, she had to ask George what the latest sound had been.

"Sounded like somebody hitting a milk bottle with a ball peen hammer," said George.

"I'd think it would be real interesting, hearing all the different sounds," said Hazel a little envious. "All the things they think up."

"Um," said George. 15

"Only, if I was Handicapper General, you know what I would do?" said Hazel. Hazel, as a matter of fact, bore a strong resemblance to the Handicapper General, a woman named Diana Moon Glampers. "If I was Diana Moon Glampers," said Hazel, "I'd have chimes on Sunday—just chimes. Kind of in honor of religion."

"I could think, if it was just chimes," said George.

"Well—maybe make 'em real loud," said Hazel. "I think I'd make a good Handicapper General."

"Good as anybody else," said George.

"Who knows better then I do what normal is?" said Hazel. 20

"Right," said George. He began to think glimmeringly about his abnormal son who was now in jail, about Harrison, but a twenty-one-gun salute in his head stopped that.

"Boy!" said Hazel, "that was a doozy, wasn't it?"

It was such a doozy that George was white and trembling, and tears stood on the rims of his red eyes. Two of of the eight ballerinas had collapsed to the studio floor, were holding their temples.

"All of a sudden you look so tired," said Hazel. "Why don't you stretch out on the sofa, so's you can rest your handicap bag on the pillows, honeybunch." She was referring to the forty-seven pounds of birdshot in a canvas bag, which was padlocked around George's neck. "Go on and rest the bag for a little while," she said. "I don't care if you're not equal to me for a while."

George weighed the bag with his hands. "I don't mind it," he said. "I don't notice it any more. It's just a part of me." 25

"You been so tired lately—kind of wore out," said Hazel. "If there was just some way we could make a little hole in the bottom of the bag, and just take out a few of them lead balls. Just a few."

"Two years in prison and two thousand dollars fine for every ball I took out," said George. "I don't call that a bargain."

"If you could just take a few out when you came home from work," said Hazel. "I mean—you don't compete with anybody around here. You just set around."

"If I tried to get away with it," said George, "then other people'd get away with it—and pretty soon we'd be right back to the dark ages again, with everybody competing against everybody else. You wouldn't like that, would you?"

"I'd hate it," said Hazel.

"There you are," said George. "The minute people start cheating on laws, what do you think happens to society?"

If Hazel hadn't been able to come up with an answer to this question, George couldn't have supplied one. A siren was going off in his head.

"Reckon it'd fall all apart," said Hazel.

"What would?" said George blankly.

"Society," said Hazel uncertainly. "Wasn't that what you just said?

"Who knows?" said George.

The television program was suddenly interrupted for a news bulletin. It wasn't clear at first as to what the bulletin was about, since the announcer, like all announcers, had a serious speech impediment. For about half a minute, and in a state of high excitement, the announcer tried to say, "Ladies and Gentlemen."

He finally gave up, handed the bulletin to a ballerina to read.

"That's all right—" Hazel said of the announcer, "he tried. That's the big thing. He tried to do the best he could with what God gave him. He should get a nice raise for trying so hard."

"Ladies and Gentlemen," said the ballerina, reading the bulletin. She must have been extraordinarily beautiful, because the mask she wore was hideous. And it was easy to see that she was the strongest and most graceful of all the dancers, for her handicap bags were as big as those worn by two-hundred pound men.

And she had to apologize at once for her voice, which was a very unfair voice for a woman to use. Her voice was a warm, luminous, timeless melody. "Excuse me—" she said, and she began again, making her voice absolutely uncompetitive.

"Harrison Bergeron, age fourteen," she said in a grackle squawk, "has just escaped from jail, where he was held on suspicion of plotting to overthrow the government. He is a genius and an athlete, is under-handicapped, and should be regarded as extremely dangerous."

A police photograph of Harrison Bergeron was flashed on the screen—upside down, then sideways, upside down again, then right side up. The picture showed the full length of Harrison against a background calibrated in feet and inches. He was exactly seven feet tall.

The rest of Harrison's appearance was Halloween and hardware. Nobody had ever born heavier handicaps. He had outgrown hindrances faster than the H-G men could think them up. Instead of a little ear radio for a mental handicap, he wore a tremendous pair of earphones, and spectacles with thick wavy lenses. The spectacles were intended to make him not only half blind, but to give him whanging headaches besides.

Scrap metal was hung all over him. Ordinarily, there was a certain symmetry, a military neatness to the handicaps issued to strong people, but Harrison looked like a walking junkyard. In the race of life, Harrison carried three hundred pounds. 45

And to offset his good looks, the H-G men required that he wear at all times a red rubber ball for a nose, keep his eyebrows shaved off, and cover his even white teeth with black caps at snaggletooth random.

"If you see this boy," said the ballerina, "do not—I repeat, do not—try to reason with him."

There was the shriek of a door being torn from its hinges.

Screams and barking cries of consternation came from the television set. The photograph of Harrison Bergeron on the screen jumped again and again, as though dancing to the tune of an earthquake.

George Bergeron correctly identified the earthquake, and well he might have—for many was the time his own home had danced to the same crashing tune. "My God—" said George, "that must be Harrison!" 50

The realization was blasted from his mind instantly by the sound of an automobile collision in his head.

When George could open his eyes again, the photograph of Harrison was gone. A living, breathing Harrison filled the screen.

Clanking, clownish, and huge, Harrison stood—in the center of the studio. The knob of the uprooted studio door was still in his hand. Ballerinas, technicians, musicians, and announcers cowered on their knees before him, expecting to die.

"I am the Emperor!" cried Harrison. "Do you hear? I am the Emperor! Everybody must do what I say at once!" He stamped his foot and the studio shook.

"Even as I stand here" he bellowed, "crippled, hobbled, sickened—I am a greater ruler than any man who ever lived! Now watch me become what I can become!" 55

Harrison tore the straps of his handicap harness like wet tissue paper, tore straps guaranteed to support five thousand pounds.

Harrison's scrap-iron handicaps crashed to the floor.

Harrison thrust his thumbs under the bar of the padlock that secured his head harness. The bar snapped like celery. Harrison smashed his headphones and spectacles against the wall.

He flung away his rubber-ball nose, revealed a man that would have awed Thor, the god of thunder.

"I shall now select my Empress!" he said, looking down on the cowering people. "Let the first woman who dares rise to her feet claim her mate and her throne!" 60

A moment passed, and then a ballerina arose, swaying like a willow.

Harrison plucked the mental handicap from her ear, snapped off her physical handicaps with marvelous delicacy. Last of all he removed her mask.

She was blindingly beautiful.

"Now—" said Harrison, taking her hand, "shall we show the people the meaning of the word dance? Music!" he commanded.

The musicians scrambled back into their chairs, and Harrison stripped them of their handicaps, too. "Play your best," he told them, "and I'll make you barons and dukes and earls." 65

The music began. It was normal at first—cheap, silly, false. But Harrison snatched two musicians from their chairs, waved them like batons as he sang the music as he wanted it played. He slammed them back into their chairs.

The music began again and was much improved.

Harrison and his Empress merely listened to the music for a while—listened gravely, as though synchronizing their heartbeats with it.

They shifted their weights to their toes.

Harrison placed his big hands on the girls tiny waist, letting her sense the weightlessness that would soon be hers. 70

And then, in an explosion of joy and grace, into the air they sprang!

Not only were the laws of the land abandoned, but the law of gravity and the laws of motion as well.

They reeled, whirled, swiveled, flounced, capered, gamboled, and spun.

They leaped like deer on the moon.

The studio ceiling was thirty feet high, but each leap brought the dancers nearer to it. 75

It became their obvious intention to kiss the ceiling. They kissed it.

And then, neutraling gravity with love and pure will, they remained suspended in air inches below the ceiling, and they kissed each other for a long, long time.

It was then that Diana Moon Glampers, the Handicapper General, came into the studio with a double-barreled ten-gauge shotgun. She fired twice, and the Emperor and the Empress were dead before they hit the floor.

Diana Moon Glampers loaded the gun again. She aimed it at the musicians and told them they had ten seconds to get their handicaps back on.

It was then that the Bergerons' television tube burned out. 80

Hazel turned to comment about the blackout to George. But George had gone out into the kitchen for a can of beer.

George came back in with the beer, paused while a handicap signal shook him up. And then he sat down again. "You been crying," he said to Hazel.

"Diana Moon Glampers loaded the gun again. She aimed it at the musicians and told them they had ten seconds to get their handicaps back on."

"Yup," she said.

"What about?" he said.

"I forget," she said. "Something real sad on television." 85

"What was it?" he said.

"It's all kind of mixed up in my mind," said Hazel.

"Forget sad things," said George.

"I always do," said Hazel.

"That's my girl," said George. He winced. There was the sound of a rivetting gun in his head. 90

"Gee—I could tell that one was a doozy," said Hazel.

"You can say that again," said George.

"Gee—" said Hazel, "I could tell that one was a doozy."

Understanding the Text

1. Why doesn't Hazel have any handicaps? Do you think she needed some? Why or why not?
2. Why doesn't George want to remove any of his weights?
3. Why is Harrison shot? Do you think anyone realizes what happened to him? Explain.

Reflection and Response

4. Why does Hazel repeat the phrase "Gee — I could tell that one was a doozy" at the end of the story?
5. Vonnegut writes that the people in his story "were equal every which way" (par. 1). Why is this kind of equality problematic?
6. Did Vonnegut write a cautionary tale about equality, or was he warning us about how technology can be misused? Or was his point about something else? Explain your response.

Making Connections

7. Does today's technology help equalize society at all? If so, is it for better or for worse? Explain.
8. George did not risk removing his weights, even for a minute, as Hazel suggested, because he was too worried the government would find out. Can you think of a parallel to real life? In other words, is there something we all avoid doing in our everyday lives for fear of someone or something catching us?
9. "Harrison Bergeron" was published in 1961, well before many of today's technologies were even an idea, let alone a reality. Vonnegut's story, however, is eerily prophetic: that is, aspects of his story seem today like they could be reality. Using evidence from "Harrison Bergeron" and another reading of your choice from this chapter, argue whether part(s) of this story could happen today.

Why Cambridge Analytica Matters

Scott Lucas

Scott Lucas is a professor of International Politics at the University of Birmingham in the United Kingdom. He is also the editor-in-chief of *EA Worldview*, a publication that aims to provide content gathered by those in the face of various news events versus looking to those in power to produce their stories' contents. Lucas's specialties lie in U.S. and British foreign policy and international relations, particularly pertaining to the Middle East and Iran. In the following article, Lucas looks at how seemingly benign social media data can used as political weapons in an effort to control public opinion.

In 1928 Edward Bernays—nephew of Sigmund Freud and father of modern advertising—wrote in his book *Propaganda*:

The conscious and intelligent manipulation of the organized habits and opinions of the masses is an important element in democratic society. Those who manipulate this unseen mechanism of society constitute an invisible government which is the true ruling power of our country. . . . We are governed, our minds are molded, our tastes formed, our ideas suggested, largely by men we have never heard of.

So what's the big deal about Cambridge Analytica using the profiles of 50 million Facebook users to devise political campaigns, including the 2016 run of Donald Trump? Isn't this just the latest incarnation of Bernays' maxim, "Vast numbers of human beings must cooperate in this manner if they are to live together as a smoothly functioning society"?

Not quite. Here's why the unfolding story of Cambridge Analytica—which tried to craft the "right" outcome in the U.S. but also in [other] countries, which vilified the "wrong" candidate in Kenya's Presidential election, and which sought the "right" result in the 2016 Brexit referendum in the U.K.—matters.

1. Size and Speed

Let's start with the obvious. In 1928, and for decades afterwards, Bernays was talking about campaigns based on collection of data from relatively small samples, taken over weeks and even months, to produce the right product or the right candidate. "Focus groups" remained focused until the waning of the twentieth century.

Cambridge Analytica, through an allegedly deceptive operation, snared 50 million accounts from Facebook in 2016. And within days,

"Cambridge Analytica, through an allegedly deceptive operation, snared 50 million accounts from Facebook in 2016."

that material was being turned by the organization and its affiliates into efforts such as the "Lock Her Up" chants against Democratic nominee Hillary Clinton, or the slogans on London buses about the European Union taking money from "our" National Health Service.

2. Possible Violation of U.S. Election Law

There is a clear provision in U.S. election legislation that foreign nationals cannot be involved in campaign-related decisions. 5

But that is precisely what U.K. national Alexander Nix, the CEO of Cambridge Analytica, set out on the camera of Britain's Channel 4, when he thought he was pitching his services to a Sri Lankan political operative: "We did all the research, all the data, all the analytics, all the targeting, we ran all the digital campaign, the television campaign and our data informed all the strategy."

Nigel Oakes, the director and cofounder of the Cambridge Analytica–affiliated SCL group; Alexander Taylor, the acting chief executive of Cambridge Analytica; Mark Turnbull, managing director of Cambridge Analytica; Christopher Wylie, who helped found the firm but is now blowing the whistle on it—all are foreign nationals.

And it's not as if the executives were unaware of the U.S. legislation. Wylie said it was a "dirty little secret" that was often discussed.

3. Entrapment and Bribery

Cambridge Analytica did not just "scrape" data and turn it into campaign themes, according to Nix and Turnbull. They boasted, in the Channel 4 videos, of schemes that included bribery and entrapment using young women to entice foreign officials.

Nix, who said the firm worked with British and Israeli spies, explained: 10

"Deep digging is interesting, but you know equally effective can be just to go and speak to the incumbents and to offer them a deal that's too good to be true and make sure that that's video recorded.

You know these sort of tactics are very effective, instantly having video evidence of corruption. . . .

Send some girls around to the candidate's house, we have lots of history of things,"

4. A Russian Link?

Cambridge Analytica's work for the Trump campaign in 2016 ran in parallel with Russian efforts, using stolen data, to assist the Republican nominee by damaging Hillary Clinton.

The Russians had their own "Lock Her Up" effort, including payments to Americans who wore Hillary Clinton masks and prison uniforms. They bought Facebook ads and promoted anti-Clinton news through social media as well as on their State outlets like RT.

There is no evidence so far that the efforts of Russia and those of Cambridge Analytica, overseen by Trump campaign manager Steve Bannon, were ever linked. But both strands followed a meeting between Trump's inner circle—Donald Trump Jr., Trump son-in-law Jared Kushner, and Bannon's predecessor Paul Manafort—and three Russians in June 2016 to discuss the provision of anti-Clinton material to the campaign.

5. And Then There's Brexit

The extent of Cambridge Analytica's data effort in the Brexit referendum has yet to be revealed. However, the firm is part of an evolving story about a payment to the Leave campaign which may have violated U.K. election laws.

In spring 2016, the BeLeave advocacy group—portrayed as "independent"—suddenly received £625,000, much of which was spent on work by Aggregate IQ, a sub-contractor for Cambridge Analytica.

Wylie and BeLeave's Shahmir Sanni say BeLeave was effectively a cover for the official Vote Leave campaign to use the money, which would have taken Vote Leave beyond the legal limit for election efforts.

Wylie told a select Parliamentary committee last week of a "common plan" to get around the law and said, "[There] could have been a different outcome had there not been, in my view, cheating."

6. Transparency

Beyond the headline-grabbing revelations is another key difference between the 1920s and 2018.

Edward Bernays was not shy about his role in data collection and manipulation. He wrote bestselling books, publicly advised major corporations, campaigns, and governments.

Until it was outed by whistle-blowers and undercover videos in recent weeks, Cambridge Analytica had not acknowledged its efforts. Nor had its clients, from Kenya to the U.S. to the U.K.

Kenyan women giving birth in the streets in 2020, a criminal Hillary Clinton, £350 million extra per week to spend on public services in a U.K. outside the EU—all this was put before us as fact and possibility, rather than the creation of a PR firm which may have been involved in illegal collection of data, disinformation, bribery, and entrapment.

All of this is an important twenty-first-century update on Bernays's 1928 observation:

"We are dominated by the relatively small number of persons . . . who understand the mental processes and social patterns of the masses. It is they who pull the wires which control the public mind."

Understanding the Text

1. What is Cambridge Analytica? What kind of data did it collect, and what happened to said data?
2. According to this article, Cambridge Analytica played a part in the Brexit vote. What does Lucas say about the company's role with the United Kingdom's politics? What does Brexit have to do with the 2016 election in the United States?

Reflection and Response

3. Lucas says that Cambridge Analytica gathered this data "through an allegedly deceptive operation" (par. 4). Is there anything in this article that evidences whether or not it was, in fact, a "deceptive operation"? Is being deceptive the same as being illegal?
4. Lucas opens his piece by quoting Edward Bernays, whom Lucas calls the father of modern advertising. Bernays says, "We are governed, our minds molded, our tastes formed, our ideas suggested, largely by men we have never heard of." Why did Lucas choose to open with this quote? How does it relate to Cambridge Analytica?

Making Connections

5. Conduct some research into *how* Cambridge Analytica gathered this data. Based on what you've learned in previous readings in this chapter, do the company's actions seem commonplace and legal? Explain using evidence from this chapter and evidence from credible sources outside of this text.
6. Lucas explains that "there is no evidence so far that the efforts of Russia and those of Cambridge Analytica, overseen by Trump campaign manager Steve Bannon, were ever linked" (par. 13). Conduct some research using credible and scholarly sources, and answer this question: Is that claim still true today? Why or why not?
7. Research and learn about Siri, Alexa, Amazon Echo, and other voice command apps and devices. Do they play a role in collecting this type of data? If not, could they in the future? How so?

Future Workplaces: Smile, You're on Camera

The Economist, a London-based publication established in 1843, is a publishing leader in topics related to the economy, which today often includes technology and its relationship with the economy. What follows is an excerpt from a special report by *The Economist* that examines how AI is being employed in various businesses. AI is offering companies new, innovative ways to assess their employees' needs, the businesses' needs, and more. While the new information has proven to be extremely useful, these new technologies also open doors to potential problems. As the article explains, the more data is collected and employed, the more companies are opening themselves up for data breaches, with potentially serious moral and legal consequences.

Walk up a set of steep stairs next to a vegan Chinese restaurant in Palo Alto in Silicon Valley, and you will see the future of work, or at least one version of it. This is the local office of Humanyze, a firm that provides "people analytics." It counts several *Fortune* 500 companies among its clients (though it will not say who they are). Its employees mill around an office full of sunlight and computers, as well as beacons that track their location and interactions. Everyone is wearing an ID badge the size of a credit card and the depth of a book of matches. It contains a microphone that picks up whether they are talking to one another; Bluetooth and infrared sensors to monitor where they are; and an accelerometer to record when they move.

"Every aspect of business is becoming more data-driven. There's no reason the people side of business shouldn't be the same," says Ben Waber, Humanyze's boss. The company's staff are treated much the same way as its clients. Data from their employees' badges are integrated with information from their e-mail and calendars to form a full picture of how they spend their time at work. Clients get to see only team-level statistics, but Humanyze's employees can look at their own data, which include metrics such as time spent with people of the same sex, activity levels and the ratio of time spent speaking versus listening.

We Can See through You

Such insights can inform corporate strategy. For example, according to Mr. Waber, firms might see that a management team is communicating only with a couple of departments and neglecting others; that certain parts of a

A woman speaks to Siri through the Apple HomePod, just one of the many "smart speakers" on the market that plays music, performs internet searches, makes and receives phone calls, and acts as an AI personal assistant. Smart speakers are equipped with multiple microphones to record their surroundings, even over loud ambient sound.
Mark Kauzlarich/Bloomberg/Getty Images

building are underused, so the space should be redesigned; that teams are given the wrong incentives; or that diversity initiatives are not working.

Hitachi, a Japanese conglomerate, sells a similar product, which it has cheerily branded a "happiness meter." Employee welfare is a particular challenge in Japan, which has a special word, *karoshi,* for death by overwork. Hitachi's algorithms infer mood levels from physical movement and pinpoint business problems that might not have been noticed before, says Kazuo Vano, Hitachi's chief scientist. For example, one manufacturing client found that when young employees spent more than an hour in a meeting, whole teams developed lower morale.

5 Employers already have vast quantities of data about their workers. "This company knows much more about me than my family does," says Leighanne Levensaler of Workday, a software firm that predicts which employees are likely to leave, among other things. Thanks to the internet, smartphones and the cloud, employers can already check who is looking at a document, when employees are working and whether they might be stealing company files and contacts. AI will go further, raising concerns about Orwellian snooping by employers on their workers. In January Amazon was granted a pair of patents for wristbands that monitor warehouse workers'

exact location and track their hand movements in real time. The technology will allow the company to gauge their employees' productivity and accuracy. JD.com, the Chinese e-commerce firm, is starting to experiment with tracking which teams and managers are the most efficient, and using algorithms to predict attrition among workers.

"It does not take much imagination to see that some companies, let alone governments, could take this information-gathering too far."

The integration of AI into the workplace will offer some benefits to workers and might even save lives. Companies with a high-risk work environment are starting to use computer vision to check whether employees are wearing appropriate safety gear, such as goggles and gloves, before giving them access to a danger area. Computer vision can also help analyze live video from cameras monitoring factory floors and work environments to detect when something is amiss. Systems like this will become as "commonplace as CCTV cameras are in shops," says Alastair Harvey of Cortexica, a firm that specializes in building them.

Employees will also be able to track their own movements. Microsoft, the software giant, already offers a program called MyAnalytics which puts together data from e-mails, calendars and so on to show employees how they spend their time, how often they are in touch with key contacts and whether they multitask too much. It also aggregates the data and offers them to managers of departments so they can see how their teams are doing. "It doesn't have that 'big brother' element. It's designed to be more productive," insists Steve Clayton of Microsoft. The idea is that individuals' data are not given out to managers, though it is not clear whether workers believe that. As part of a broader investment in AI, Microsoft is also starting to use the technology to translate the monthly question-and-answer session held by the company's boss, Satya Nadella, for its workers worldwide, and analyze employees' reactions.

It does not take much imagination to see that some companies, let alone governments, could take this information-gathering too far. Veriato, an American firm, makes software that registers everything that happens on an employee's computer. It can search for signals that may indicate poor productivity and malicious activity (like stealing company records), and scans e-mails to understand how sentiment changes over time. As voice-enabled speakers become more commonplace at work, they can be used to gather ever more data.

This is of particular concern in authoritarian states. In China increasing numbers of firms, and even some cities, use cameras to identify employees for the purpose of giving them access to buildings. More troubling, the government is planning to compile a "social credit" score for all its citizens, pooling online data about them to predict their future behavior.

All this may require a new type of agreement between employers and employees. Most employment contracts in America give employers blanket rights to monitor employees and collect data about them, but few workers are aware of that. Mr. Waber of Humanyze thinks these data should have better legal protection, especially in America (Europe has stronger privacy laws). 10

As more companies rely on outside firms to collect and crunch employee information, privacy concerns will increase, and employees may feel violated if they do not think they have given their consent to sharing their data. Laszlo Bock, who used to run Google's human-resources department and now heads a start-up focused on work, reckons that "it's going to play out in a bad way before it plays out in a good way."

Understanding the Text

1. What are "people analytics," according to the reading?
2. What is AI? What is its purpose? Feel free to use Google to help you better understand AI.
3. How is AI helping businesses in today's market, according to the reading?

Reflection and Response

4. The article says that "AI will go further, raising concerns about Orwellian snooping by employers on their workers" (par. 5). What is the term "Orwellian snooping" referencing? Should employers have unadulterated access to their employees while they are working? Why or why not?
5. The article examines a new technology Amazon is testing out, which allows Amazon to track its "employees' productivity and accuracy." What are the benefits of this technology? What are some potential pitfalls or problems this may cause?

Making Connections

6. Can you connect the notion of "Orwellian Snooping" to an idea or ideas present in "Harrison Bergeron?" Why or why not? Be sure to back up your response with textual evidence from both readings.
7. Using the internet, learn more about China's "social credit" score (par. 9). Does this make you support AI more or less? Defend your stance with scholarly and credible research.
8. It is clear, based on the reading, that AI offers a lot of benefits to companies, and can even benefit individual employees. However, the end of the piece touches on the potential danger of companies taking their use of AI too far. Write an essay that argues for or against the use of AI in the workplace. Your essay must cover the opposing argument and argue why your stance is more logical.

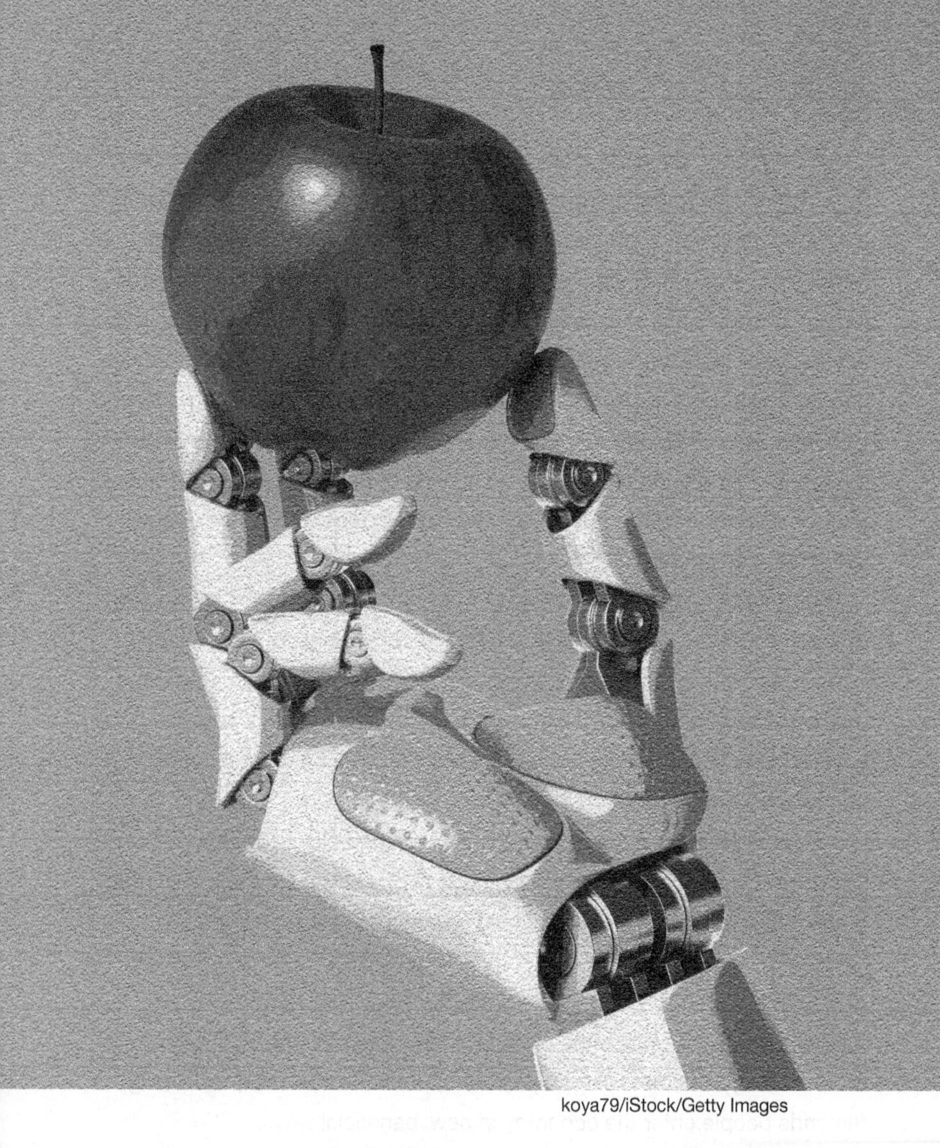

koya79/iStock/Getty Images

4 Who Controls the Economy, Us or the Technology We've Created?

What if someone told you that future retailers would line their aisles with gigantic vending machines, machines that could dispense small couches, mini refrigerators, computers, and maybe even a car? Sounds odd and too futuristic to happen, right? Actually, some economists and IT gurus argue the above scenario will in fact happen sooner rather than later. As technology advances, so does our way of life. In today's world, the tech industry is booming. Students graduating college with degrees in STEM earn more, on average, than other majors fresh out of college, and for a good reason: the world we live in needs people to run the machines we increasingly rely on — but what happens when the machines no longer need people?

This chapter explores the future we may soon be faced with, a machine-operated economy, and the reality we're already seeing: the loss of jobs to robot labor. Lisa Fickenscher, in her piece "The Cutthroat Jobs Strategy Amazon Uses to Conquer Retail," argues that Amazon and its ability to sell almost all products is the reason why 50,000 jobs were lost in a particular service sector. In "Drones Go to Work," by Chris Anderson, the author details the various ways drones are being employed in a myriad of industries. Even robots, once simply the stuff of science fiction, are starting to affect everyday jobs. "Robots Could Replace 1.7 Million American Truckers in the Next Decade," by Natalie Kitroeff, looks at how driverless trucks could potentially end the need for truckers. These authors worry that technology will reduce the need for many people in the workforce, but technology, at times, can do just the opposite: sometimes, technology demands people enter the economy in new, beneficial ways.

Matt Britton, in "The Peer-to-Peer Economy," looks at how technology is allowing companies and individuals to work together in new ways. eBay, Facebook Marketplace, and Craigslist are just a few of the platforms that allow individuals to infiltrate the economy, giving them the chance to earn a living in ways never before seen — at least not on such a large,

photo: koya79/iStock/Getty Images

worldwide scale. In "Sweden's Push to Get Rid of Cash Has Some Saying, 'Not So Fast'," Liz Alderman expands on Britton's topic, investigating how technology is (almost) eliminating the need for cash. This does not, however, mean technology and the economy always work hand-in-hand, nor does it mean that experts fully understand the impact of technology on the economy.

For example, Bitcoin, a fairly new monetary system, which Ole Bjerg explains in "How Is Bitcoin Money?," is drastically changing the way people view conventional money. Bitcoin is 100 percent electronic and it solely relies on computers, eliminating all tangible aspects of currency. Sure, many of us who use "real" money never actually *see* cash. Our paychecks are directly deposited into our bank, and we use credit and debit cards to make purchases and pay bills, and we just assume our money is where it should be: in the bank. But still, any one of us could go to a bank and withdraw actual, tangible cash should we want to. Not with Bitcoin. Bitcoin is allowing today's citizens to circumvent mainstream monetary procedures, allowing people to complete transactions outside the government, stimulating the economy in an ultramodern way. How long it will last and how well it will work remains to be seen, but it does reveal technology's ability to affect our ever-growing, ever-changing, tech-dependent economy.

As you read this chapter, think about the ways you have seen technology affect the economy just in the last few years. Think, too, about generations that may benefit from this economy more than others. Meagan Johnson's piece, "Stop Talking about Work/Life Balance!," explains how technology is working for millennials in many ways and how it manages to hinder them, too. Our relationship with technology and the economy is ever evolving. From Uber to drones, robots to Bitcoin, technology (and the people who develop it) finds ways to affect the economy — positively, negatively, and enigmatically.

The Cutthroat Jobs Strategy Amazon Uses to Conquer Retail

Lisa Fickenscher

Lisa Fickenscher, who double-majored in English and German as an undergraduate at Guilford College, has worked in journalism for over three decades, contributing to a number of publications and news outlets, including Fox News, *MarketWatch*, and *Crain's New York Business*, and has received many awards for her "online scoops." Currently, she is a retail reporter for the *New York Post* where she covers consumer product companies.

In the following article, Fickenscher explores the ubiquity of Amazon and its various economic effects — both positive and negative. She challenges Amazon's recent job creation claims, arguing that the company may not be creating as many jobs as it would like the public to believe and may even be cutting more jobs than it creates.

Jonathan and Michele Rhudy recently moved their Richmond, Va., consulting business into new digs and bought all new stuff—office supplies, furniture and even a coffee maker.

There was no question in their minds where all of those items, plus paper clips, reams of paper, and every last doodad, would come from: Amazon.

Jonathan said he discovered the magic of Jeff Bezos's e-commerce giant almost a decade ago while he and his wife, both busy running their small business, were raising their three kids and were always running out of diapers.

"I quickly realized how, being a family of five, we didn't have time to go to the store," he told *The Post* this week.

5 The Rhudys' buying habits mimic those of millions of Americans, and that surely makes Bezos very happy—not to mention immensely wealthy.

Bezos, who has grown his humble, money-losing online book shop in 1995 into one of the most powerful economic engines on earth—selling everything from airplane parts to zebra-print dresses—saw his net worth grow alongside his company. This week, he became the world's second-richest person, worth $76 billion.

Amazon, which is also in the movie-producing business—it won an Oscar for its "Manchester by the Sea"—is now the fourth-most-valuable company in the United States and employs 341,400 full- and part-time employees.

Amazon will add another 100,000 full-time jobs over the next 18 months, Bezos proudly announced this year.

The new jobs are great, but a closer inspection shows Amazon may simply be adding back jobs it helped kill off.

For example, Amazon played a large role in eliminating more than 50,000 jobs in recent years from just three companies—Staples, Office Depot, and Best Buy, public filings show. 10

In March, MarketWatch estimated that Amazon will destroy 1.5 million retail jobs in the next five years. And with its push into self-driving trucks, drone delivery, automated grocery stores and more, the site said the total number of lost jobs would likely be more than 2 million, concluding, "Could Amazon actually kill more American jobs than China did? It's quite likely."

In addition to capturing 50 cents of every dollar spent online, Amazon, according to a report by the Institute for Local Self-Reliance, a nonprofit research organization, is used for half the online shopping searches undertaken by U.S. consumers.

"With such control over retail sales, jobs, ad dollars, and more, critics are beginning to wonder if Amazon is good for America."

In other words, Bezos knows what you want to buy before you buy it.

With such control over retail sales, jobs, ad dollars, and more, critics are beginning to wonder if Amazon is good for America.

One thing is for sure, Bezos isn't ready to slow down—he wants Amazon to control even more of Americans' lifestyles. 15

A report several weeks ago from Bloomberg revealed that Amazon has invited executives from companies that make major consumer brands—like Nike, Oreos, and Cheerios—to a meeting in May where the company hopes to convince them not to sell their products through Walmart and other bigbox retailers.

Bezos wants the brands to be sold direct to consumers through Amazon.

The company is also spreading its wings deeper into media with its $50 million deal last week to stream ten Thursday-night NFL games next season. Bezos paid five times what Twitter paid for the streaming rights in 2016.

There is also chatter Amazon will soon add freighters to its retail arsenal, which already includes trucks, plans for drones and some planes.

Not everyone is scared of Amazon's growing influence and power. 20

"Retail always evolves and reflects society, and right now, consumers are getting more value for their money," said Richard Kestenbaum, a partner in Triangle Capital. "That makes our society stronger and it forces other retailers to be more creative and competitive."

But more and more are casting a wary eye on the Seattle company, whose brown boxes with the "smiling" arrows are ubiquitous.

Critics say Amazon is crushing local jobs and tax bases.

"People need to have jobs to be able to afford online delivery, and Amazon has knocked out so many private-sector retail jobs and will knock out public-sector jobs as well," said Burt Flickinger, managing director of Strategic Resource Group, a retail consulting firm.

For years, there has been talk about Amazon ruining better-paying retail jobs and replacing them with lower-paying ones. It has tried to fight back—but there is still a strong belief that jobs at Bezos's company don't pay well.

"This is a company that is so sophisticated in its use of the web and supply chain, but it adopts this retrograde way of employing labor," said Nelson Lichtenstein, director of the Center for the Study of Work, Labor and Democracy at the University of California, Santa Barbara.

"Here are these 500 to 800 workers toiling away in a big warehouse, but many aren't employed directly by Amazon, and if they have a complaint, Amazon says, 'It's not our problem,' because they are contract workers."

Most of the jobs Amazon creates are low-paying warehouse and customer-service positions. The warehouse jobs pay, on average, $12.32 an hour, 9 percent less than the industry average at other warehouses, according to ILSR, which looked at 1,300 Amazon wage postings on Glassdoor.com.

Recent Amazon ads for customer-service reps who work from home show that the company is paying just $10 an hour—or just $2.75 more than the federal minimum wage.

Although Amazon defends its salaries, touting that it's 30 percent higher than what other retailers pay, critics say the comparison is not apples to apples.

Bezos disagrees.

"These jobs are not just in our Seattle headquarters or in Silicon Valley—they're in our customer-service network, fulfillment centers, and other facilities in local communities throughout the country," he said in a statement.

But research shows that Amazon's job gains have come at the expense of other jobs and that the quality of the Amazon positions is inferior, with about 40 percent of the workforce in its warehouses considered contract or temporary employees, according to ILSR.

For years, one of Amazon's biggest edges over competitors was that it didn't collect state sales tax, giving it a price advantage equaling as much as 9 percent in states like New York. That income, experts say, deprived states of revenue and contributed to budget shortfalls.

That changed on April 1. Amazon is now collecting state taxes in the forty-five states that have them, according to the Tax Policy Center.

"But the damage was done," said Flickinger. "The money that was not collected can't be made up, and those budget deficits are going to force state governments to lay off employees at unprecedented rates."

Bezos is hardly cowed by critics or allegations Amazon is a job wrecker.

Such taunts almost seem to embolden the brash entrepreneur, whose vision for the company includes convenience stores without cashiers, deliveries with drones and, irony of ironies, opening bookstores.

Bezos has shown an unlimited ability to spend money to accomplish his goal of dominating different sectors.

And now, the grocery business may be in his crosshairs. He is developing an Amazon Go store, which will allow the company's Prime customers to pull items off shelves and carry them home without passing through a cashier's line or opening their wallets. 40

Customers simply scan their phones on a kiosk when they enter the store, and Amazon technology charges their account after they leave the store.

The stores require way fewer employees—and could drive rival grocers out of business.

Bezos's hiring announcements generate excitement in the media, most recently in January, when he revealed a plan to add 100,000 employees over the following eighteen months.

But left unsaid are all the jobs quietly killed off over the years—in bookstores, electronics chains, office-supply shops and soon, perhaps, supermarkets.

And maybe, too, another block in another town left a little less active. 45

Understanding the Text

1. According to this article, is Amazon actually creating new jobs? Why or why not?
2. Jeff Bezos, unsurprisingly, disagrees with many, if not all, of the negative claims about Amazon. How does Bezos refute the idea that Amazon is doing more harm than good?

Reflection and Response

3. "With such control over retail sales, jobs, ad dollars, and more, critics are beginning to wonder if Amazon is good for America" (par. 14), writes Fickenscher. Identify ways Amazon both benefits and harms the American economy. Be sure to back up your claims with textual evidence.
4. Fickenscher says that Amazon is in talks with big names such as Oreo and Nike in an effort to see if Amazon can gain exclusive rights to sell their products. Based on the information presented in the reading, do you think Amazon's proposal would be good for the economy? Why or why not?

5. Is Fickenscher's tone positive or negative throughout her piece? Does this tone help or hurt the piece's ability to garner your support? Explain your response using textual evidence.

Making Connections

6. Locate and read the article "The Amazon Effect and Online Shopping Are Not Killing Retail" by Laura Myers, published by *CMS Connected* on June 22, 2018. Compare Myers's argument with Fickenscher's. Who makes a stronger argument about what is currently happening in retail, Fickenscher or Myers? How so? Be sure to support your response with textual evidence from both readings.
7. Research some of Bezos's current philanthropic endeavors. Do his charitable efforts negate issues presented in this article? Why or why not?

Drones Go to Work

Chris Anderson

Chris Anderson has been a writer for over three decades. He wrote for *The Economist* for seven years, covering technology and business. In 2004, his article "The Long Tail" was published in *WIRED Magazine* and was later expanded into a book focusing on particular retail strategies affected by technology. It is no surprise, then, that Anderson's economic and technology expertise continued to expand, and in 2009, he founded 3D Robotics, a company that manufactures drones.

In this piece from *Harvard Business Review*, where Anderson is a regular contributor, he explains how drones have infiltrated various areas of business, helping everyone from construction workers to farmers. He does a clear job illustrating the solid financial benefits companies reap when they embrace drone usage: time and money are easily saved. But at what cost? As you read the essay, think about the negative effects ubiquitous drone use may cause.

Every morning at the construction site down the street from my office, the day starts with a familiar hum. It's the sound of the regular drone scan, when a small black quadcopter flies itself over the site in perfect lines, as if on rails. The buzz overhead is now so familiar that workers no longer look up as the aircraft does its work. It's just part of the job, as unremarkable as the crane that shares the air above the site. In the sheer normalness of this—a flying robot turned into just another piece of construction equipment—lies the real revolution.

"Reality capture"—the process of digitizing the physical world by scanning it inside and out, from the ground and the air—has finally matured into a technology that's transforming business. You can see it in small ways in Google Maps, where data is captured by satellites, airplanes, and cars, and presented in both 2-D and 3-D. Now that kind of mapping, initially designed for humans, is done at much higher resolution in preparation for the self-driving car, which needs highly detailed 3-D maps of cities in order to efficiently navigate. The methods of creating such models of the real world are related to the technology of "motion capture," which drives movies and video games today. Normally that requires bringing the production to the scanners—putting people in a large room outfitted for scanning and then creating the scene. But drones flip that, allowing us to bring the scanner to the scene. They're just regular cameras (and some smart software) precisely revolving around objects to create photo-realistic digital models.

In some ways it's astonishing that we're using drones on construction sites and in movies. Ten years ago the technology was still in labs. Five years ago it was merely very expensive. Today you can buy a drone at

Walmart that can do real enterprise work, using software in the cloud. Now that it's so cheap and easy to put cameras in the sky, it's becoming commercially useful. Beyond construction, drone data is used in agriculture (crop mapping), energy (solar and wind turbine monitoring), insurance (roof scanning), infrastructure (inspection), communications, and countless other industries that touch the physical world. We know that "you can manage only what you can measure," but usually measuring the real world is hard. Drones make it much easier.

"The drone economy is real, and you need a strategy for exploiting it."

Industries have long sought data from above, generally through satellites or planes, but drones are better "sensors in the sky" than both. They gather higher-resolution and more-frequent data than satellites (whose view is obscured by clouds over two-thirds of the planet at any time), and they're cheaper, easier, and safer than planes. Drones can provide "anytime, anywhere" access to overhead views with an accuracy that rivals laser scanning—and they're just getting started. In this century's project to extend the internet to the physical world, drones are the path to the third dimension—up. They are, in short, the "internet of flying things."

You might think of drones as toys or flying cameras for the GoPro set, and that is still the lion's share of the business. But like the smartphone and other examples of the "commercialization of enterprise" before them, drones are now being outfitted with business-grade software and becoming serious data-collection platforms—hardware as open and extensible as a smartphone, with virtually limitless app potential. As in any app economy, surprising and ingenious uses will emerge that we haven't even thought of yet; and predictable and powerful apps will improve over time.

Or you might think of drones as delivery vehicles, since that's the application—consumer delivery—that the media grabs on to most ferociously when seeking click-generating amazing/scary visions of the future. Frankly, delivery is one of the least compelling, most complicated applications for drones (anything that involves autonomously flying in crowded environments is the black-diamond slope of technology and regulation). Most of the industry is focused on the other side of the continuum: on data, not delivery—commercial use over privately owned land, where the usual concerns about privacy, annoyance, and scary robots overhead are minimized.

Drone economics are classically disruptive. Already drones can accomplish in hours tasks that take people days. They can provide deeply detailed visual data for a tiny fraction of the cost of acquiring the same data by other means. They're becoming crucial in workplace safety, removing people from precarious processes such as cell-tower inspection. And they offer, literally, a new view into business: Their low-overhead perspective is bringing new insights and capabilities to fields and factories alike.

Like any robot, a drone can be autonomous, which means breaking the link between pilot and aircraft. Regulations today require that drones have an "operator" on the ground (even if the operation is just pushing a button on a smartphone and idly watching as the drone does its work). But as drones are getting smarter, regulators are starting to consider flights beyond "visual line of sight"—ones in which onboard sensors and machine vision will more than compensate for the eyes of a human on the ground far away. Once such fully autonomous use is allowed, the historic "one pilot/one aircraft" calculus can become "one operator/many vehicles" or even "*no* operator/many vehicles." That's where the real economic potential of autonomy will kick in: When the marginal cost of scanning the world approaches zero (because robots, not people, are doing the work), we'll do a lot more of it. Call this the "democratization of earth observation": a low-cost, high-resolution alternative to satellites. Anytime, anywhere access to the skies.

The drone economy is real, and you need a strategy for exploiting it. Here's how to think about what's happening—and what's going to happen. We'll start back at the construction site, a work environment in desperate need of what drones can provide.

Capturing Reality for the Cost of a Nice Lunch

The construction industry is the world's second largest (after agriculture), worth $8 trillion a year. But it's remarkably inefficient. The typical commercial construction project runs 80% over budget and twenty months behind schedule, according to McKinsey. 10

On-screen, in the architect's CAD file, everything looks perfect. But on-site, in the mud and dust, things are different. And the difference between concept and reality is where about $3 trillion of that $8 trillion gets lost, in a cascade of change orders, rework, and schedule slips.

Drones are meant to close that gap. The one buzzing outside my window, taking passes at the site, is capturing images with a high-performance camera mounted on a precision gimbal. It's taking regular photos (albeit at very high resolution), which are sent to the cloud and, using photogrammetry techniques to derive geometries from visual data, are turned into photo-realistic 2-D and 3-D models. (Google does the same thing in Google Maps, at lower resolution and with data that might be two or three years old. To see this, switch to Google Earth view and click on the "3-D" button.) In the construction site trailer, the drone's data shows up by mid-morning as an overhead view of the site, which can be zoomed in for detail the size of a U.S. quarter or rotated at any angle, like a video game or virtual reality scene. Superimposed on the scans are the CAD files used to guide the construction—an "as designed" view overlaid on an "as built" view. It's like an augmented reality lens into what should be versus what *is,* and the difference between the two can be worth thousands of dollars a day in cost savings on each site—billions across the industry. So the site superintendent monitors progress daily.

Mistakes, changes, and surprises are unavoidable whenever idealized designs meet the real world. But they can be minimized by spotting clashes early enough to fix them, work around them, or at least update the CAD model to reflect changes for future work. There are lots of ways to measure a construction site, ranging from tape measures and clipboards to lasers, high-precision GPS, and even X-rays. But they all cost money and take time, so they're not used often, at least not over the entire site. With drones, a whole site can be mapped daily, in high detail, for as little as $25 a day.

Rising from the Ground to Fill the Missing Middle

The ascent of the drone economy is a steep one. Ten years ago unmanned aerial vehicles were military technology, costing millions of dollars and cloaked in secrecy. But then came the smartphone, bringing with it a suite of component technologies, from sensors and fast processors to cameras, broadband wireless, and GPS. All those chips enabled the remarkable supercomputer in your pocket, but the economies of scale of smartphone production also made them cheap and available for other uses. The first step was to transform adjacent industries, including robotics. I call this proliferation of components "the peace dividend of the smartphone wars."

Companies including my own came out of this moment. Cheap high-powered components and a maker's attitude allowed enthusiasts and entrepreneurs to reimagine drones not as coming down from higher in the sky but as rising from the ground. Rather than seeing "airplanes without pilots," we saw "smartphones with propellers." Moving at the pace of the smartphone industry, not the aerospace industry, drones went from hackers' devices to hobbyists' instruments to toys costing less than $100 at your local big-box store in less than four years—perhaps the fastest transfer of technology from CIA to Costco in history. Five years ago the main commercial objection to the word "drone" was that it had military connotations. Now it's that people think of the aircraft as playthings. Has any word changed its meaning from "weapon" to "toy" faster?

And it doesn't end there. Wave one was technology, wave two was toys, and now comes the third and most important wave. Drones are becoming *tools*. The market for people who want flying selfie cameras may be limited, but the market for data about the physical world is as big as the world itself.

Drones are starting to fill the "missing middle" between satellites and street level, digitizing the planet in high resolution and near-real time at a tiny fraction of the cost of alternatives.

The trajectory of this third wave—drones as tools—is more dramatic than that of the two preceding waves. First drones will populate the skies in increasing numbers as regulations and technology allow safer use. Estimates vary widely; some data predicts that by next year more than 100,000 operators will be managing 200,000 drones that will fill the sky, doing some work or another.

Next, the market for drone apps will explode as more and more people find ingenious uses. Drones will remain primarily data-collection vehicles, but the breadth of apps for them is only just beginning to be discovered. For example, drones have already been used for search and rescue and for wildlife monitoring. They can provide wireless internet access (something Facebook is investing in) and deliver medicine in the developing world. And they can not only map crops but also spray them with pesticides or deposit new seeds and beneficial insects.

Then, drones will gain even greater cost advantages when they don't just remove the pilot from the cockpit but remove the pilot entirely. The true breakthrough will come with autonomy. 20

Autonomous, Small, and Countless

Technology to allow drones to fly themselves exists and is improving quickly, going from simple GPS guidance to true visual navigation—the way a human would fly. Take humans out of the loop, and suddenly aircraft look more like the birds that inspired them: autonomous, small, and countless; born for the air and able to navigate it tirelessly and effortlessly. We are as yet tourists in the air, briefly visiting it at great cost. By breaking the link between man and machine, we can occupy the skies. The third dimension is the last frontier on Earth to be properly colonized (yes, both up to the skies and down under the seas, but we'll leave the latter to our aquatic-drone cousins). Colonize it we will, but as with space and the ocean depths, we'll use robots, not humans.

Why now? A combination of three trends. First, the price/performance bounty of the smartphone tech we talked about earlier made drones cheap and good. For example, the gyroscopic and other sensors packed into a tiny $3 chip in your phone were just a decade ago mechanical devices costing as much as $100,000 and mounted in enclosures ranging in size from lunch boxes to dorm fridges.

Second, the ability to make cheap and good drones put them within the reach of regular consumers (willing to spend up to $1,000) who had a real use case (aerial video and photography). As a result, companies had to make them easy to use—just swipe and fly—to drive adoption. Drones had to become more sophisticated as users became *less* sophisticated.

Third, once the consumer drone boom unexpectedly put more than a million drones—ranging from small toys to high-end "prosumer" models—into the skies over America in less than four years under a "recreational use" exemption to the FAA's strict rules about flying things, the regulators had to respond. To steer the market toward safer use without inhibiting it, the agency accelerated rules that would allow drones to be used commercially without the need for pilots' licenses or special waivers. The new rules took effect in August 2016, essentially kicking off the commercial drone era.

Cool Is Not Enough

Where all this really kicks in is the enterprise. There, nobody is using a drone because it's cool. They're using it because it does a job better than the alternative. All that matters is the job, and every step that stands between wanting the job done and having it done is friction that inhibits adoption. The perfect enterprise drone is a box with a red button. When you push the button, you get your data. Anything more complicated is a pain point to be eliminated. (And after that, we'll get rid of the button, too.)

What that means is seamless integration between drones and enterprise software, such that all the data is automatically collected, sent to the cloud, analyzed, and displayed in useful form, ideally in near—real time.

What will this look like? Although it might surprise you, I hope the future of drones is boring. As the CEO of a drone company, I obviously stand to gain from the rise of drones, but I don't see that happening if we are focused on the *excitement* of drones. The sign of a successful technology is not that it thrills but that it becomes essential and accepted, fading into the wallpaper of modernity. Electricity was once a magic trick, but now it is assumed. The internet is going the same way. My end goal is for drones to be thought of as just another unsexy industrial tool, like agricultural machinery or generators on construction sites—as obviously useful as they are unremarkable.

My inspiration in this is my grandfather, Fred Hauser, who in the 1930s invented the automatic sprinkler system (his patents decorate our walls). You may not think of a sprinkler system as a robot, but it is: Today's are connected to the internet, collect data, operate autonomously, and, best of all, just work. Now imagine farm drones doing the same: boxes scattered around the farm with copters inside and solar cells outside, to recharge their batteries. Like the irrigation systems, at some point in the day they wake up, emerge from the boxes, and do their thing—crop mapping, pest spotting, or even fertilizing like bees. When they're done, they return automatically to their boxes; the lids close, and they sleep until they do it all again the next day. All the farmer needs to know is that the daily crop report on his or her phone is extraordinarily detailed, with multispectral analysis of everything from disease to dampness, measured to the individual leaf and analyzed by machine-learning software to flag issues and make recommendations for the day's work.

Drones as ubiquitous as sprinklers: We've come a long way from weapons, sci-fi movies, and headlines. But in the prosaic° applications of advanced technologies lie their real impact. Once we find drones no longer novel enough to be worthy of HBR articles, my work will be done.

Prosaic: something commonplace or ordinary.

Understanding the Text

1. How does the author define "reality capture"?
2. Anderson says that "drone economics are classically disruptive" (par. 7). What does that mean? What are drones doing to the economy?
3. Why does the author say to focus more on the apps than the drone itself? Why are the apps more important?

Reflection and Response

4. Anderson says "reality capture" is necessary for self-driving cars; that is, self-driving cars need detailed pictures of the roads they will drive on. Do you foresee any potential problems relying on these pictures for self-driving cars? Why or why not?
5. In the article's last paragraph, Anderson says that his future hope is that drones will be considered prosaic one day, as commonplace as automatic sprinklers. Why? If that happens, what do you predict that will mean for the economy? Explain your reasoning with evidence from this reading or other readings from the chapter.

Making Connections

6. Drones are "becoming serious data-collection platforms," according to Anderson (par. 5). Using evidence from this reading and at least one reading from Chapter 3, answer this question: is drone use disconcerting at all? Why or why not? If you said yes, do the benefits of drone use outweigh the potential risks? Explain your stance.
7. Anderson's piece sustains an upbeat, hopeful tone, while focusing on the positive effects of the ever-growing ubiquitous nature of drones. Write an essay challenging his positive claims about the benefits of drone use.

Robots Could Replace 1.7 Million American Truckers in the Next Decade

Natalie Kitroeff

Natalie Kitroeff is a reporter for the *New York Times*, covering industry and the economy. Since graduating from Princeton University, Kitroeff has contributed to a number of notable publications such as the *Los Angeles Times* and *Bloomberg Businessweek*, writing on a whole spectrum of subjects, including court cases, student debt, the economy, higher education, and more.

The following piece by Kitroeff was published in the *Los Angeles Times* in 2016. Here, she examines how driverless cars may be the end of the trucking industry as we know it. Automated vehicles would significantly lower the cost of transporting goods, making it a desirable goal, but Kitroeff suggests the reach of self-driving cars may not end with the trucking industry.

Trucking paid for Scott Spindola to take a road trip down the coast of Spain, climb halfway up Machu Picchu, and sample a Costa Rican beach for two weeks. The 44-year-old from Covina now makes up to $70,000 per year, with overtime, hauling goods from the port of Long Beach. He has full medical coverage and plans to drive until he retires.

But in a decade, his big rig may not have any need for him.

Carmaking giants and ride-sharing upstarts racing to put autonomous vehicles on the road are dead set on replacing drivers, and that includes truckers. Trucks without human hands at the wheel could be on American roads within a decade, say analysts and industry executives.

At risk is one of the most common jobs in many states, and one of the last remaining careers that offer middle-class pay to those without a college degree. There are 1.7 million truckers in America, and another 1.7 million drivers of taxis, buses and delivery vehicles. That compares with 4.1 million construction workers.

5 While factory jobs have gushed out of the country over the last decade, trucking has grown and pay has risen. Truckers make $42,500 per year on average, putting them firmly in the middle class.

On Sept. 20, the Obama administration put its weight behind automated driving, for the first time releasing federal guidelines for the systems. About a dozen states already created laws that allow for the testing of self-driving vehicles. But the federal government, through the National Highway Traffic Safety Administration, will ultimately have to set rules to safely accommodate 80,000-pound autonomous trucks on U.S. highways.

In doing so, the feds have placed a bet that driverless cars and trucks will save lives. But autonomous big rigs, taxis and Ubers also promise to lower the cost of travel and transporting goods.

"It would also be the first time that machines take direct aim at an entire class of blue-collar work in America."

It would also be the first time that machines take direct aim at an entire class of blue-collar work in America. Other workers who do things you may think cannot be done by robots—like gardeners, home builders and trash collectors—may be next.

"We are going to see a wave and an acceleration in automation, and it will affect job markets," said Jerry Kaplan, a Stanford lecturer and the author of "Humans Need Not Apply" and "Artificial Intelligence: What Everyone Needs to Know," two books that chronicle the effect of robotics on labor.

"Long-haul truck driving is a great example, where there isn't much judgment involved and it's a fairly controlled environment," Kaplan said. 10

Robots' march into vehicles, factories, stores, and offices could also profoundly deepen inequality. Research has shown that artificial intelligence helps erase jobs that require basic skills and creates more roles for highly educated people.

"Automation tends to replace low-wage jobs with high-wage jobs," said James Bessen, a lecturer at the Boston University School of Law who researches the effect of innovation on labor.

"The people whose skills become obsolete are low-wage workers, and to the extent that it's difficult for them to acquire new skills, it affects inequality."

Trucking will likely be the first type of driving to be fully automated—meaning there's no one at the wheel. One reason is that long-haul big rigs spend most of their time on highways, which are the easiest roads to navigate without human intervention.

But there's also a sweeter financial incentive for automating trucks. 15 Trucking is a $700-billion industry, in which a third of costs go to compensating drivers.

"If you can get rid of the drivers, those people are out of jobs, but the cost of moving all those goods goes down significantly," Kaplan said.

The companies pioneering these new technologies have tried to sell cost savings as something that will be good for trucking employers and workers.

Otto, a self-driving truck company started by former Google engineers and executives, pitches its system as a source of new income for drivers who will be able to spend more time in vehicles that can drive solo as they rest.

Here, a pizzaiolo robot makes a pizza, showing how robotics is quickly infiltrating various aspects of the economy.
Philippe Wojazer/Reuters

Uber bought the San Francisco-based company in August.

The start-up retrofits trucks with kits allowing them to navigate freeways without a driver actually holding the wheel. For the last several months, at least one Volvo truck equipped with the software has been test driving, with a person at the wheel, on Interstate 280 or on the 101 Freeway in California. 20

The system works by installing a set of motion sensors; cameras; lidar, which uses laser light; and computer software to make driving decisions.

Lior Ron, the company's cofounder, says that as the system gathers data on tens of thousands of miles of U.S. highways, having the driver asleep in the back could become a possibility within the next few years. That would instantly double the amount of time a truck spends on the road per day, allowing freight companies to charge more for shorter delivery times, Ron said. "The truck can now move 24/7."

Ron says that the question of whether to pay drivers for hours spent sleeping in a truck while it drives for them has "been an ongoing debate in the trucking industry."

Otto says its system may eventually allow some big rigs to traverse highways without a driver at all. In that scenario, a truck driver would drive the big rigs to and from "pick up and drop off locations," playing a role "similar to a tug boat," but trucks could drive without any human present during the longest stretches of the journey, says Ron, the cofounder.

Several states are already laying the groundwork for a future with fewer truckers. In September, the Michigan state senate approved a law allowing trucks to drive autonomously in "platoons," where two or more big rigs drive together and synchronize their movements. That bill follows laws passed in California, Florida, and Utah that set regulations for testing truck platoons.

Wirelessly connected trucks made their European debut in April, when trucks from six major carmakers successfully drove in platoons through Sweden, Germany, Belgium, and the Netherlands.

Josh Switkes, 36, says those convoys will be on American roads within a year. Switkes is the chief executive of Peloton, a Mountain View-based company whose software links two semi-trailer trucks. Peloton's investors include UPS and Volvo Group. The company has begun taking reservations for its system from freight fleets, and it plans to start delivering them "in volume" within a year.

The system works by transmitting very specific data from the first truck to the second truck so they operate in tandem, almost as if they were a train on the open road. When the lead truck brakes, the following vehicle receives a signal telling it how much to apply its brake.

This EZ-PRO vehicle, which debuted in France in 2018, can drive on its own. It even turns into a pop-up store or a food truck, if needed.

Philippe Wojazer/Reuters

That close communication can make it safe for the trucks to drive as close as thirty feet from one another, the company says. If a car cuts in between the two, the rear truck can automatically slow down, assume a safe following distance, and then return to its previous arrangement with the leader after the car changes lanes, Switkes said.

Reduced drag on the first and second vehicles can produce massive fuel savings. For the time being, drivers are installed in both trucks, with their feet off the brakes and accelerators, and their hands on the wheel. 30

Peloton says its technology reduces fuel expenses by 7 percent and could save companies even more on salaries.

"As we move to higher levels of automation, we can save them massive amounts in labor [costs]," said Switkes. He said that Peloton could make the rear truck in the convoy fully machine-driven, without any humans present, within a decade.

Even before that happens, though, platooning could segregate drivers into different pay classes depending on whether they're driving the first or second rig.

"Maybe you pay the front driver more because they have a more important job," said Switkes. Eventually, as the system makes trucks more efficient, "you may be paying fewer drivers overall," he added.

Like most companies trying to turn trucks into robots, Peloton sees itself being useful mainly on highways, which are more predictable and less people-filled than city streets. Still, the company announced in July that it will develop and help deploy technology to power a fleet of heavy-duty trucks to serve the San Diego port. 35

Truck driver Spindola, perched atop a creaky seat as his big rig sat inside the Long Beach port waiting to be OKd for a departure to a nearby storage yard, said he isn't convinced that a machine could ever do his job.

"You need a human being to deal with some of the problems we have out on the road," Spindola said. There are too many delicate maneuvers involved, he maintained, too many tricks and turns and unforeseen circumstances to hand the wheel over to a robot.

He had just spent about twenty minutes weaving in and out of corridors of 40-foot shipping containers at the port to attach a chassis to his rig and drive it toward a large forklift. There the lift operator slowly slotted a container onto the bay of his truck.

"This is a good job as far as pay, one of the last good jobs," Spindola said. "Maybe I just don't want to accept that the future is here."

Understanding the Text

1. Why are driverless cars a desirable goal? Support your answer with at least three points from this reading.
2. Why does Kitroeff argue that trucking would be the first type of driving to become fully automated? Do you agree with her reasoning? Why or why not?

Reflection and Response

3. Kitroeff quotes Jerry Kaplan, a lecturer at Stanford University who believes that an uptick in automation is coming and is poised to affect the job market. Is there a chance the uptick would have a positive effect? Why or why not? Back up your answer with textual evidence from the reading.
4. This reading tells us that driverless trucks would better serve the environment than the trucks we see on the road today. Does that mean it is okay to eliminate 1.7 million jobs? Why or why not?

Making Connections

5. Spindola, a truck driver, argued that road hazards are better handled by people behind the wheel, implying driverless cars would not be able to handle the "delicate maneuvers" drivers face. What do you think? Conduct brief research into driverless cars and trucks. Based on what you learn, do you feel driverless trucks are inevitable, or is Spindola right? Be sure to support your answer with research.
6. Kitroeff warns, "Other workers who do things you may think cannot be done by robots – like gardeners, home builders and trash collectors – may be next" (par. 8). Research some of the latest technologies, looking for ones specifically replacing tasks currently performed by humans. Based on your research, write an essay arguing whether or not Kitroeff is correct in saying that those jobs, and others, will soon be at risk.
7. Kitroeff argues driverless cars would mean 1.7 million truckers would lose their jobs. Think back to Lisa Fickenscher's article, "The Cutthroat Jobs Strategy Amazon Uses to Conquer Retail" (p. 198). According to Fickenscher, Amazon has eliminated jobs, too, but Amazon claims those jobs were simply replaced. Would the same happen here? In other words, would driverless trucks cost people their jobs but replace them in a new way? You may need to conduct brief research to answer this thoroughly.

The Peer-to-Peer Economy; from *YouthNation: Building Remarkable Brands in a Youth-Driving Culture*

Matt Britton

Self-proclaimed millennial expert Matt Britton has spent over twenty-five years helping businesses better integrate their brands with their consumers. Now, Britton is the founder and CEO of a marketing intelligence platform called Suzy, which allows businesses to gather real-time feedback from audiences regarding business decisions. Suzy also works with research and development teams, creative campaigns, and advertisers to better help and target their goal audiences.

Britton's book, *YouthNation* (2015), explores youth as a commodity rather than an age. *YouthNation* gives companies tangible ways to better connect with today's hyper-socialized world. In this excerpt, Britton looks at how the economy allows peers to network in ways that have never been possible until today. While reading, think about the benefits and potential drawbacks of a peer-to-peer economy.

Just like in the old village square where we relied on our neighbors and fellow villagers to supply what we could not provide for ourselves, we are witnessing a community-fueled renaissance° of sharing among what just a decade ago would be seen as complete strangers. The urbanization° of America and its newfound population density has spawned a new powerful by-product of the internet: the peer-to-peer economy.

When eBay first exploded in the late nineties during the original internet boom, the novelty of selling your old stuff directly to somebody else without the need for a middleman was almost too cool to pass up. The internet's connective powers began to allow consumers from all walks of life to buy and sell anything directly to each other. A user-generated ratings system helped ensure you wouldn't get ripped off, fostering a thriving community of eBay users. Today eBay is a $60 billion company, and it stands tall among a growing graveyard of traditional retailers like Circuit City and Blockbuster Video.

Renaissance: a rebirth of an activity.
Urbanization: the process of making a rural place more urban in lifestyles and population.

Why buy a camera from a store when there is an unused one sitting in a closet ten blocks away at a fraction of the price? Why pay for a dog sitter tonight when someone three doors down with a love for dogs and an empty calendar is willing to do it for free, or for a service you can provide in exchange? Increasingly the products and services YouthNation is looking for are coming from peers instead of businesses, and as a result an entirely different outlook for capitalism and consumption is emerging. In fact, *Fast Company* magazine estimates that the peer-to-peer economy has an estimated value of over $25 billion.[1]

The 3 P's of P2P

The peer-to-peer economy has three core pillars that sustain it, each willing and able catalysts to this disruptive movement.

PROVIDERS: Individuals with products to sell or services to offer, which can benefit others through price, convenience, accessibility, and choice

PARTAKERS: Willing participants in the peer-to-peer economy who are buying and trading products and services

PLATFORMS: The technologies, almost all of which are based on mobile applications, which facilitate the peer-to-peer transactions

Source: Jeremiah Owyang

Bartering Is Back

Not only has the peer-to-peer economy generated massive transactional volume between peers, but it has also created a renaissance in the bartering space. Consumers are now increasingly bartering with one another without any dollars exchanging hands. Whether its product-for-product, service-for-service, or product-for-service, the bartering space is a serious throwback to simpler times that is suddenly new and relevant again.

Yerdel, for example, is a modern-day thrift shop for the peer-to-peer age. Participating users donate used durable goods to Yerdel in exchange for credits. Those credits are then used to purchase other people's used goods. We all have storage closets full of gently used stuff, and for YouthNation this presents the perfect opportunity to bypass traditional and costly retail channels to get their hands on the stuff that they truly want.

The Peer-to-Peer Payment Revolution

When money is actually switching hands YouthNation is gravitating toward a new class of payment tools that allow for quick and easy electronic transactions between individuals. Venmo, for instance, is an easy-to-use application allowing consumers to send and receive money with anyone in their phone's contact list. All you need to do is open the app, choose your recipient, and the money will appear in your recipient's Venmo account instantly. Venmo is also being embraced by YouthNation when it's time to split the bill at a restaurant or buy festival tickets. 10

Given the explosion of the peer-to-peer economy, companies big and small are getting into the game. Google's payments product "Google Wallet" now allows consumers to pay one another as well as for retail purchases. In late 2014, the popular messaging startup Snapchat announced the launch of Snapcash, a product destined to gain traction within YouthNation. In spring 2015 Facebook announced free peer-to-peer payment integration into its Messenger product.

The Currency of Trust: Ratings and Reviews

A critical component of the peer-to-peer economy is trust. Without it there would be a lack of governance and the floodgates for fraud and deceit would be opened. What if the person you are buying from is a fraud or the product is flawed? Originally brought to the mainstream by eBay, user ratings from both buyers and sellers have become the linchpin of trust for the peer-to-peer economy. Today, when you are finished with a ride in an Uber car, both the driver and rider are rated by each other. When you are finished with a stay at an Airbnb residence, both the landlord and tenant rate each other. Over time the ratings of both buyer and seller create a trusted user identity. The more positive ratings that are received, the more confidence is built into the business transaction.

"In fact, *Fast Company* magazine estimates that the peer-to-peer economy has an estimated value of over $25 billion."

User ratings have today indeed become the FICO score of the peer-to-peer economy. If one wants to participate, one needs to earn the trust of the community, and it takes time to both build and repair your peer-to-peer credit score. Companies are also slowly starting to take notice of the importance of highly rated peer-to-peer economy providers. In late 2014, Toyota partnered with DiscoLyft, one of the most highly rated and prolific drivers on Lyft (an Uber competitor) for an advertising campaign.

Peer-to-Peer Disruption

As consumers gravitate toward one another to transact and disintermediate traditional providers, there is suddenly a lot at stake for corporate America. As YouthNation reprioritizes their spending toward experiences and away from durable goods, the economic benefits and convenience of peer-to-peer are starting to disrupt every industry in its path. A recent study by UC Berkeley revealed that one shared car in the marketplace can result in lost auto sales of over $270,000.[2] This doesn't even take into account lost revenue in parking, loans fees, taxes, gas, etc.

The question for brands has now become "What role does my company play in the lives of youth, given that they are increasingly looking to transact with one another?" As corporate America begins to reinvent itself to compete in a YouthNation-driven marketplace with such a transformed set of priorities, it will have to grapple with the fact that its future consumers prefer access to ownership. In valuing the status update over the status symbol, the proven model of selling shiny new things at top retail rates may one day become obsolete.

Several companies of late have taken notice and successfully created partnerships or investments to integrate their business model into the peer-to-peer economy.

Peer-to-Peer Standout Models

- In 2014 Ikea partnered with Airbnb to allow consumers to stay overnight in their Sydney, Australia, location.
- Ford now offers discounts on its Explorer model to Uber and Lyft drivers and is now offering customized vehicles including built-in USB chargers.
- In 2014, eleven companies announced they were building products to integrate with Uber's API (or product platform). This allows companies like Starbucks and Hyatt to integrate directly into Uber's app for discounts and easy navigation to their locations.

While partnering with companies that are embracing peer-to-peer models certainly signals to consumers that a brand is embracing change, businesses must recognize that they are in a whole new world and disrupt themselves accordingly because YouthNation simply is not the same audience as previous generations.

The peer-to-peer economy will create opportunities for massive winners and epic losers much like the digital media transformation has forced in the past decade. Here are some examples of companies that are embracing the future.

Build It: Rent the Runway

Nestled in the second floor corridor of the red-hot Cosmopolitan Resort among traditional luxury retailers in Las Vegas is a prime location of a relatively little known internet and brick-and-mortar brand called "Rent the Runway," a classic "why didn't I think of that" idea. The concept is brilliant. Searching for that perfect dress to wear out to the neighboring Marquee nightclub, women can visit a Rent the Runway retail location or online platform, pick a dress, accessorize it, and rent the whole ensemble for the night. For a fraction of the cost, shoppers can feel and look stylish for their memorable moment. There are no commitments to buy, and shoppers can use the opportunity to experiment with new styles.

This of course comes at the expense of the neighboring boutiques that traditionally have benefitted from shoppers purchasing elaborate outfits, which they would likely wear once and store in a closet forever. Rent the Runway creates a clear win/win. For the business, renting out clothing is a highly lucrative model that eliminates the traditional retail complexities of inventory management. For consumers, they are now offered an affordable and fun way to experiment with fashion at a time when they are dying to do so, without breaking the bank. 20

Buy It: Avis and ZipCar

Despite the fact that Avis and other major rental car companies like Hertz technically already play in the rental and sharing game, they have become victims of their dated business models as of late. The cumbersome process of renting a car, coupled with the exorbitant fee structure and inaccessible rental locations, have put them at risk in an increasingly YouthNation-driven culture and economy.

Enter ZipCar. Founded way back in 1999, ZipCar is a true innovator that saw the future of transportation long before Uber was a verb in tech circles. With hubs in twenty-six American cities and over one million members worldwide, ZipCar users tap into a mobile app to locate cars that are strategically parked in highly accessible city locations, which can be rented for a flat hourly fee. Using a card (no keys required), drivers can enter and start cars and drive right away. The rental fees include gas, taxes, and insurance and allow users a carefree way to access cars on a short-term basis.

Faced with an uncertain future, Avis acquired ZipCar for $500 million in 2013. In doing so, Avis went the route that Blockbuster did not when given the chance to buy Netflix. They acknowledged the tides of change in their industry and bought their way into a future model that would have likely otherwise brought about their demise.

Extend It: Coca-Cola's Wonolo

Faced with the tedious challenge of restocking shelves, coolers, displays, and vending machines and amidst a growing landscape of youth unemployment, Coca-Cola ventured outside of the beverage business to launch Wonolo. The purpose of this venture is to create short-term work opportunities by providing finite tasks within a particular locale.

For Coca-Cola, Wonolo provides an immediate business benefit in the creation of an effective tool to accomplish the arduous tasks involved in running the world's largest beverage company. On a more long-term basis, though, Coke has made the necessary step in diving into the sharing economy by extending its business to meet the needs of YouthNation head-on.

As YouthNation continues its migration into cities the ability to meet, transact, and yes, even date will become faster and easier through the power of peer-to-peer dynamics. The peer-to-peer economy is now a foundation of YouthNation lifestyle, and stands to disrupt every business in its path in the years ahead.

Notes

1. Amy Kamenetz, "Why the Sharing Economy Is Growing," *Fast Company* magazine, 2013 (http://www.fastcoexist.com/1682080/why-the-sharing-economy-is-growing).
2. Elliot Martin and Susan Shaheen, "The Impact of Carsharing on Household Vehicle Ownership," *Access, The Magazine of UCTC*, 2011 (http://www.uctc.net/access/38/access38_carsharing_ownership.shtml).

Understanding the Text

1. What is a peer-to-peer economy? How is it different from previous economies?
2. Why is the internet responsible for our current peer-to-peer economy?
3. According to Britton, why is trust vital to a peer-to-peer economy?

Reflection and Response

4. Bartering, often thought of as an archaic, obsolete method of payment, is back, according to Britton. Why? What are the benefits of modern-day bartering?
5. Britton explains how the peer-to-peer economy relies on user ratings to maintain the vital trust needed to keep it going. Have you ever left or received user ratings online for a job or transaction? What are the potential pitfalls of relying on user ratings?

Making Connections

6. Britton shared how a recent University of California, Berkeley, study illustrated how just one shared car in the marketplace can mean losing $270,000 in auto sales. Does this mean ride-sharing companies likes Uber and Lyft actually have a negative effect on the economy? Explain your answer using textual evidence from this reading and at least one other reading from this chapter.
7. Early readings in this chapter emphasize how technology and machines are replacing people. How does this piece challenge that idea? Be sure to pull textual evidence from this reading and earlier readings from the chapter.
8. Watch Martin Ford's TED Talk called "How We'll Earn Money in a Future without Jobs." In his TED Talk, he explains that humans have always feared technology replacing jobs and/or older technologies; he references cars replacing the horse and buggy, but he says it is different this time; today's machines, according to Ford, are nothing like the cars that replaced those buggies. Why, according to Ford, is the present-day situation different from those of the past?

Sweden's Push to Get Rid of Cash Has Some Saying, 'Not So Fast'

Liz Alderman

After graduating from the University of Virginia, Liz Alderman joined the ranks at *Bridge News*, where she reported on United States economics. Later, Alderman began reporting for the *International Herald Tribune*, where she was a business editor focused on international business and economics. Alderman is based in Paris and currently serves as chief European business correspondent for the *New York Times*, which is where the following piece originally appeared.

In this article, Alderman explains Sweden's move toward a cashless society, meaning they are utilizing technology that can take the place of physical money. While many young people are embracing this change, Alderman raises some areas of concern.

Few countries have been moving toward a cashless society as fast as Sweden. But cash is being squeezed out so quickly—with half the nation's retailers predicting they will stop accepting bills before 2025—that the government is recalculating the societal costs of a cash-free future.

The financial authorities, who once embraced the trend, are asking banks to keep peddling notes and coins until the government can figure out what going cash-free means for young and old consumers. The central bank, which predicts cash may fade from Sweden, is testing a digital currency—an e-krona—to keep firm control of the money supply. Lawmakers are exploring the fate of online payments and bank accounts if an electrical grid fails or servers are thwarted by power failures, hackers or even war.

"When you are where we are, it would be wrong to sit back with our arms crossed, doing nothing, and then just take note of the fact that cash has disappeared," said Stefan Ingves, the governor of Sweden's central bank, known as the Riksbank. "You can't turn back time, but you do have to find a way to deal with change."

Ask most people in Sweden how often they pay with cash, and the answer is "almost never." A fifth of Swedes, in a country of 10 million people, do not use automated teller machines anymore. More than 4,000 Swedes have implanted microchips in their hands, allowing them to pay for rail travel and food, or enter keyless offices, with a wave. Restaurants, buses, parking lots and even pay toilets depend on clicks rather than cash.

Consumer groups say the shift leaves many retirees—a third of all Swedes are 55 or older—as well as some immigrants and people with disabilities at a disadvantage. They cannot easily gain access to electronic means for some goods and transactions, and rely on banks and their customer service. 5

And the progress toward a cashless society could upend the state's centuries-old role as sovereign guarantor.° If cash disappears, commercial banks would wield greater control.

"We need to pause and think about whether this is good or bad, and not just sit back and let it happen," said Mats Dillén, the head of a Swedish Parliament committee studying the matter. "If cash disappears, that would be a big change, with major implications for society and the economy."

"More than 4,000 Swedes have implanted microchips in their hands, allowing them to pay for rail travel and food, or enter keyless offices, with a wave."

Urban consumers worldwide are increasingly paying with apps and plastic. In China and other Asian countries rife with young smartphone users, mobile payments are routine. In Europe, about one in five people say they rarely carry money. In Belgium, Denmark and Norway, debit and credit card use has hit record highs.

But Sweden—and particularly its young people—is at the vanguard. Bills and coins represent just 1 percent of the economy, compared with 10 percent in Europe and 8 percent in the United States. About one in ten consumers paid for something in cash this year, down from 40 percent in 2010. Most merchants in Sweden still accept notes and coins, but their ranks are thinning.

Among 18- to 24-year-olds, the numbers are startling: Up to 95 percent of their purchases are with a debit card or a smartphone app called Swish, a payment system set up by Sweden's biggest banks. 10

Ikea, whose flat-box furniture is a staple of young households, has been experimenting to gauge the allure and effect of cashless commerce. In Gavle, about 100 miles north of Stockholm, managers decided to go cashless temporarily last month after they realized that fewer than 1 percent of shoppers used cash—and Ikea employees were spending about 15 percent of their time handling, counting and storing money.

Patric Burstein, a senior manager, said the cashless test had freed employees to work on the sales floor. So far, around 1.2 of every 1,000 customers have been unable to pay with anything but cash—and mainly in the cafeteria, where people tend to spend change. Rather than bother with bills, Ikea has been offering those customers freebies.

Guarantor: someone or something that guarantees something.

"We said, 'If you want a 50 cent hot dog, be my guest, take it. But next time maybe you can bring a card,'" said Mr. Burstein, who is 38.

The test so far suggests that cash is not essential and, instead, may be costly, he said. "We're spending a lot of resources on a very small percentage that actually need the service," he said.

15 The nearby branch of the Swedish National Pensioners Organization has led protests against the experiment, in part, because many retirees like to go to the Gavle Ikea for a bite to eat.

"We have around one million people who aren't comfortable using the computer, iPads or iPhones for banking," said Christina Tallberg, 75, the group's national president. "We aren't against the digital movement, but we think it's going a bit too fast."

The organization has been raising money to teach retirees how to pay electronically, but, paradoxically, that good effort has been tripped up by an abundance of cash. When collections for training are taken in rural areas—and the seniors donate in cash—the pensioner in charge must drive miles to find a bank that will actually take the money, Ms. Tallberg said. About half of Sweden's 1,400 bank branches no longer accept cash deposits.

"It's more or less impossible, because the banks refuse to take cash," she said.

Banks have propelled the cashless revolution by encouraging consumers and retailers to use debit and credit cards, which yields banks and credit card companies lucrative fees. That includes the bank-developed Swish smartphone app.

20 Sweden's banks have cut back on cash in part for safety reasons after a rash of violent robberies in the mid-2000s. The national psyche is marked by an infamous helicopter heist in Vastberga in 2009, when thieves landed on the roof of a G4S cash service depot and stole millions—a drama now being turned into a Netflix film. Last year, only two banks were robbed, compared with 210 in 2008.

In recent years, banks have dismantled cash machines by the hundreds. So little cash is used now that it has become expensive to track and maintain, said Leif Trogen, an official at the Swedish Bankers' Association.

There are two proposals by the Swedish authorities to keep cash at hand. Parliament wants just the biggest banks to handle cash. The central bank is holding out for all banks to keep money flowing. Swedbank, SEB and other big Swedish financial institutions are fighting the lawmakers' demands, saying it would place an undue burden on them to provide greater access.

"The demand for cash is decreasing at an ever faster pace," Mr. Trogen said. "Therefore, it is fundamentally wrong to legislate to influence the demand for cash."

The central bank has plans to roll out a pilot version next year of a new type of Riksbank money—the digital krona, or e-krona—that could replace physical cash or at least help calm the current cash conundrum. An e-krona would mean that the functions of a currency backed by the state would remain, even in an all-digital world that is fast approaching.

Christine Lagarde, managing director of the International Monetary Fund, noted last week that several central banks were "seriously considering" digital currencies. "While the case for digital currency is not universal, we should investigate it further—seriously, carefully and creatively," she said. 25

Mr. Ingves, the central bank governor, said, "This is not a war on cash, but no one has argued that this evolutionary motion is going to stop."

Understanding the Text

1. While Alderman does not give an explicit definition, how does the article define going *cashless*?
2. Why is the Swedish government preparing for a cashless society?
3. According to the reading, companies that accept cash are losing money. Why is that?

Reflection and Response

4. Why do you think retirees struggle with Sweden's cashless move? Do you think the same is happening here in the United States? Why or why not?
5. Based on this reading, what pros and cons do you see to a cashless society? Do you see any pros and cons not addressed in the reading? If so, what are they?
6. Review the steps the Swedish government is taking to deal with the depletion of cash. Do you feel they are excited about this evolution or simply dealing with it out of necessity? Be sure to back up your response with evidence from the reading.

Making Connections

7. If cash continues to phase itself out in Sweden, Alderman argues commercial banks would wield greater control (par. 6). Research the Swedish banking system. Do you foresee an issue if commercial banks gain greater control? Why or why not? Be sure to back up your response with credible research.
8. Alderman's piece makes going cashless seem convenient for various reasons. Locate the article "Cashing Out" by Jeff Blyskal online and read it. He, too, argues convenience as vital to preferred payment options. Based on this reading, Blyskal's piece, and the rest of this chapter, what other ways does it seem consumers prioritize convenience in their lives?
9. Based on this reading and Blyskal's piece referenced in Question 8, do you think cash could completely go away? Why or why not?

How Is Bitcoin Money?

Ole Bjerg

Bitcoin is a digital currency that emerged in 2009. Because it is digital, specialized encryption techniques are used to monitor how the "money" is spent. Each user's bitcoin is stored in a type of digital wallet that keeps track of the funds. Bitcoin operates completely outside of a regular bank, which opens up many questions and concerns about this type of currency.

Ole Bjerg, an associate professor at the Copenhagen Business School in Denmark and an economics expert, centers his research on the relationship between economics and philosophy. His book, *Making Money: The Philosophy of Crisis Capitalism* (2014), looks at capitalism from a philosophical standpoint, so it is no surprise that the following paper does something similar.

Forbes Magazine declared 2013 to be "The Year of the Bitcoin" (Christensen, 2013).

(How) Is Bitcoin Money?

A straightforward way of approaching Bitcoin might have been to start with a definition of money and then analyze how Bitcoin fits the bill. This is precisely how the Bank of England has structured its recent report on digital currencies:

From the perspective of economic theory, whether a digital currency may be considered to be money depends on the extent to which it acts as a store of value, a medium of exchange and a unit of account. (Ali et al., 2014: 276)

While such analysis of Bitcoin may shed some light on the first aim of our article, it fails to accommodate the second aim as it takes the conventional definition of money as well as the conventional forms of money for granted. In contrast to such a "perspective of economic theory," we shall be pursuing a philosophical approach that takes as its premise that the very ontological° foundation of money is inherently undecidable. We are thus aligned with the position pointedly formulated by Graeber: "money has no essence. It's not 'really' anything; therefore, its nature has always been and presumably always will be a matter of political contention" (Graeber, 2011: 372). As demonstrated by Ingham, "the mainstream, or orthodox, tradition of modern economics does not attach much theoretical importance to money" (Ingham, 2004: 7). Perhaps not surprisingly, the BoE approach to Bitcoin perfectly represents such orthodox thinking,

Ontological: the philosophical study of being or existence.

where the very nature of money is rarely if ever questioned. Asking the question: how is Bitcoin money?, we thus deliberately sidestep the much more obvious question: is Bitcoin money? This in turn allows us to explore Bitcoin without first committing to any definite theory of money.

> "So what is Bitcoin? Is it the currency of the future, or is it merely just a Ponzi scheme for the internet age?"

How Does Bitcoin Work?

Bitcoin is a virtual network that allows users to transfer digital coins to each other. Each bitcoin consists of a unique chain of digital signatures that is stored in a digital wallet installed on the user's computer. The wallet generates keys used for sending and receiving coins. A transfer of bitcoins is made as the current owner of the coin uses a private digital key to approve of the addition of the recipient's key to a string of previous transactions. The coin is then transferred and now appears in the recipient's wallet with a recorded history of transactions, including the one just recently completed.

Since physical objects can only be in one place at any time, a physical coin cannot be spent simultaneously on two or more separate transactions. Once the coin is in the hands of the payee, the payer cannot spend the same coin again. A fundamental property of digital entities is, however, that they are easily copied and multiplied. In other words, digital entities can be in several places at the same time. Therefore, digital currencies are faced with the problem of double-spending. 5

Bitcoin's original solution to the problem of double-spending is what makes it fundamentally different from conventional electronic payment systems and vastly more successful than comparable predecessors (Barber et al., 2012). Rather than instituting a central authority of verification, Bitcoin is organized as a decentralized peer-to-peer network. In brief, the solution to the problem of double-spending is to keep a complete and public record of all transactions in the network. Here is how the enigmatic founder of Bitcoin, Satoshi Nakamoto, reflects on the problem and its solution:

We need a way for the payee to know that the previous owners did not sign any earlier transactions. For our purposes, the earliest transaction is the one that counts, so we don't care about later attempts to double-spend. The only way to confirm the absence of a transaction is to be aware of all transactions. In the mint based model, the mint was aware of all transactions and decided which arrived first. To accomplish this without a trusted party, transactions must be publicly announced . . . and we need a system for participants to agree on a single history of the order in which they were received. (Nakamoto, 2008: 2)

A transfer of Bitcoin is recorded with a time stamp by the network and bundled together with other transactions to form a so-called block. The block is processed by users making their CPU power available to the network. For reasons to be explained later, different users compete against each other to see who is able to process the block faster. When a block of transactions has been processed and verified, it is sealed through an operation that connects it to the previous block and it is now added to the so-called block chain. The block chain constitutes the entire history of payments in the system against which new transactions are checked for double-spending. The block chain is a public ledger of transactions and balances in the system.

Bitcoins obviously differ from these conventional moneys created in the zone of indistinction between official government and private banking enterprise, since Bitcoin is not only a system for the creation of new money but even also an independent currency. When people hear about Bitcoin for the first time, they are often puzzled by the way that the system seems to create new money out of nothing. While there are indeed reasons to be puzzled about Bitcoin, it is perhaps even more curious that we readily accept the creation of new money out of nothing by the conventional commercial bank system. At least Bitcoin is not parasitic on the national currency of any sovereign state. If Bitcoin should eventually collapse in a spiral of hyperinflation, it is only going to affect those money users who have voluntarily chosen to invest some of their money in Bitcoin. When sizeable commercial banks collapse, they tend to take with them their host organism, which is the whole currency of the state in which they are operating. This is of course why the hosting government often does not allow these parasites to fail.

Another difference between Bitcoin and conventional bank credit money is that new bitcoins are created and introduced into the economy free of debt. When commercial banks issue new credit that circulates as money, this is typically the other side of the customer's debt to the bank. Since the customer's debt to the bank is typically charged with a much higher interest rate than the bank's debt to the customer, there is an inherent tendency in this system for the total amount of debt to grow faster than the total supply of money (Binswanger, 2013). As we have already touched upon, Bitcoin is similar to state-issued fiat money in this respect. A Bitcoin does not represent a claim on any particular debtor but rather a claim upon the whole "society" of Bitcoin users. Bitcoin is credit money without debt.

It is not unlikely that in the future we are going to see credit institutions offering loans denominated in Bitcoin. Such institutions are, however, necessarily going to differ from the way that commercial banks operate within the conventional money system. In the Bitcoin system, there is only one kind of money and payments are almost immediately cleared through the block chain. This means that a commercial bank wanting to lend bitcoins to a customer would need to have this money in advance. It might borrow the money from another customer with a surplus of bitcoins or it might take it out of its own reserves. The bank does not, however, have the power to create new bitcoins by simply crediting the borrower's deposit account with the bank.

So what is Bitcoin? Is it the currency of the future, or is it merely just a Ponzi scheme for the internet age? Of course this is the question that everyone with any interest in Bitcoin is curious to find out. But perhaps this is the wrong way of posing the question of Bitcoin—at least if we believe that we can deduce the answer from the structure and constitution of Bitcoin itself. As we have seen from the preceding analysis, there is no gold, state, or debt backing the value of Bitcoin. However, this should not lead us to the conclusion that Bitcoin is in some sense a fake form of money. Or at least, Bitcoin is no more fake than more conventional forms of money.

References

Ali R, Barrdear J, Clews R and Southgate J (2014) The economics of digital currencies. *Bank of England Quarterly Bulletin*, Q3. Available at: http://www.bankofengland.co.uk/publications/Documents/quarterlybulletin/2014/qb14q3digitalcurrenciesbitcoin1.pdf (accessed 9 November 2015).

Barber S, Boyen X, Shi E and Uzun E (2012) Bitter to better: How to make Bitcoin a better currency. In: Keromytis AD (ed.) *Financial Cryptography and Data Security*. Berlin: Springer, pp. 399–414.

Binswanger HC (2013) *Die Wachstumsspirale: Geld, Energie und Imagination in der Dynamik des Marktprozesses*. Marburg: Metropolis-Verlag.

Christensen N (2013) 2013: Year of the Bitcoin. Available at: http://www.forbes.com/sites/kitconews/2013/12/10/2013-year-of-the-bitcoin/ (accessed 6 November 2015).

Graeber D (2011) *Debt: The First 5,000 Years*. New York: Melville House.

Ingham G (2004) *The Nature of Money*. Cambridge: Polity.

Nakamoto S (2008) Bitcoin: A peer-to-peer electronic cash system. Available at: https://bitcoin.org/bitcoin.pdf (accessed 6 November 2015).

Žižek S (1999) Human rights and its discontents. Lecture at Bard College, 16 November.

Understanding the Text

1. What is Bitcoin? Explain in your own words? Where in the article does Bjerg define the term most clearly in the reading?
2. What is a "block chain" (par. 7), and why is it vital to Bitcoin's success?
3. How is Bitcoin similar to gold, according to Bjerg?

Reflection and Response

3. Bjerg says that he hopes we all start to question what we "perceive as conventional money" in his video posted to Theory, Culture & Society's *YouTube* page entitled "Ole Bjerg on How Is Bitcoin Money?" Based on this reading, do you think Bitcoin has a chance at becoming more widely used and a more conventional currency? Why or why not?
4. The United States used to follow the gold standard; that is, the government could only print money equal to the amount of tangible gold our nation possessed. In 1971, however, the United States officially abandoned the gold standard, and currently, our government can print money when it wants or needs to without any limiting factor. Is Bitcoin different from today's U.S. dollar? Why or why not?
5. Are there regulations regarding how much Bitcoin exists, and why? Explain your answer with textual evidence from the reading.

Making Connections

6. Google "fiat money" and "the gold standard." How is Bitcoin different from these two concepts? Which one is it closest to? Explain, referencing this article and other scholarly sources to back up your explanation.
7. Bjerg, in the same *YouTube* video referred to in Question 3 above, argues that Bitcoin is a mixture of "commodity theory," "state theory" (i.e., "Chartalist approach"), and "credit theory," but is not all of any one of those theories. Google each theory. How is Bitcoin a little of each? Is Bitcoin its own form of money, one we have never seen? Why or why not?

Stop Talking about Work/Life Balance! TEQ and the Millennial Generation

Meagan Johnson

Meagan Johnson is a self-proclaimed generational expert. After she repeatedly encountered workplace environments rife with negative comments about Generation Xers (she worked for Quaker Oats and Kraft Foods in the 1990s), Johnson sought to prove those comments wrong. She has written about the generational differences between baby boomers (birth years between 1945 and 1964) and Generation Xers (born 1965 to 1980) for many years. Johnson now travels the country speaking to audiences about how to manage Generation Xers in the workforce, using the differences between them and baby boomers to engender positive outcomes.

In the following article, Johnson explains how the millennial generation is different from previous ones, illustrating how archaic management rules often stifle millennials' abilities to work in the best ways they know. Johnson offers simple solutions that she believes will help bridge the generational gaps in the workplace.

The millennial generation is filling the ranks of our ever-changing workforce. Sometimes referred to as Generation Y, or the Echo Boom, the millennial generation is the 80 million people born between 1980 and 2000.[1]

Mighty baby boomers take note; your millennial children have dwarfed your 72 million and are poised to become the largest generation in the workforce. Roughly 10,000 new millennials reach the legal drinking age every year (21 years old)[2] and in five years, 40 percent of the workforce will be part of this generation.[3] It is estimated that in ten years, 75 percent of the workforce across the globe will call themselves the millennial generation.[4]

The millennials are the first generation to reach adulthood during the early years of the new millennium. Millennials are more ethnically diverse than past generations, more racially accepting, more likely to have a tattoo, more open to immigration than older generations, and are more likely to sleep with their cell phones.[5]

As baby boomers' numbers in the workforce begin to diminish, it becomes more important than ever to an organization's financial success to harness the skillsets that these young people can contribute. Their managers claim this generation wants instant gratification, praise, a fun work environment and a casual dress code. Human Resources is often left feeling that millennials hold all the cards: if they do not give in to their demands, this generation will quit in the blink of an eye.

This is not the case. Millennials do want to make a difference in the workplace, and they want to challenge the "old ways of thinking." In Deloitte's survey of 7,800 young people, two of the fundamental findings were that the millennials perceive that innovation is being stifled by management's attitude, and that a company should be judged not only on a monetary basis but also on the impact it makes in society.[6] In the 2014 Millennial Impact Report, over 90 percent of millennials are aligning their skills with organizations that are making a positive impact on society.[7] They are eager to change the landscape of today's workplace and leave their individual mark. They just want your help to do it!

Here are some actions you can take to work more effectively with millennials and harness the incredible energy and fresh perspectives they bring to our workplace today.

Stop Talking About Work/Life Balance!

According to baby boomer company owner Alec Johnston, "When I hear someone tell me they want *work/life* balance, they are really telling me that they want more time off to goof around."

BusinessDictionary.com defines work/life balance (WLB) as, "A comfortable state of equilibrium achieved between an employee's primary priorities of their employment position and their private lifestyle."[8] Work/life balance is nothing new. We have all struggled with how much time we spend at the office versus our own personal interests.

Baby Boomers Live to Work

They will likely cancel personal activities if the work is not done or their boss asked them to stay late. To baby boomers, the term "work ethic" translates to putting in long hours at the office and sacrificing personal interests until the job is complete. The line between work and play is very distinct. There is time to work and time to relax.

Generation X entered the workforce in the late 1980s and early 1990s, during the recession preceded by Black Monday, the largest one day stock market crash. Gen Xer's cries for WLB became a backlash to baby boomers' intense dedication to employers and long hours spent at the office. For Generation X, WLB became synonymous with "working to live" rather than "living to work." Gen X championed the home office and began demanding flexible working hours. They felt at ease working earlier or working later to get the job done so they would have free time to spend with friends and family. The line between working and free time became hazy; they felt comfortable putting a load of laundry in, working from home in their pajamas, and making one or two appearances at the office.

What Happened to WLB and the Millennials

The millennial generation has been on the forefront of the ever-changing technology wave. They may not have invented many of the technologies we use today (think Bill Gates and Steve Jobs) but they adapted the new technologies into their lives at an early age and for many of us, have become the "experts" we turn to when our smartphones, tablets and laptops confuse us.

Millennials perceive that technology makes their lives easier and makes their generation distinctive. According to Nielsen, more than 70 percent of millennials feel that technology makes their lives simpler; more than 50 percent feel that technology brings family and friends closer; and almost 25 percent ranked technology as the top defining characteristic of their generation.[10] Technology has given the millennial generation the ability to work where they want and when they want.

It is not about balancing how much time you work versus how much time you spend on your leisure activities, it is a combination of both. It is a lifestyle of both working and personal activities that are merged into one. This is *Technology Equilibrium* (TEQ), a term I coined that describes the successful blending of life and work via technology.

> "It is estimated that in ten years, 75 percent of the workforce across the globe will call themselves the millennial generation."

Technology Equilibrium has erased the line between work and personal life altogether. The millennials do not recognize a distinction between work and personal life because technology has made it so easy to move from one to the next. According to the Cisco Connected World Technology Report, 90 percent of millennials check their phones, texts, and social media before getting out of bed and over 65 percent spend as much time if not more with their friends online versus in person.[11]

In that time before getting out of bed, the millennial has, in a typical case, responded to a customer request, communicated with his or her overseas team, updated his or her online calendar, read his or her Twitter feed and made happy hour plans with friends via OpenTable. Can Generation X and baby boomers do this? Of course they can, but for the millennials, adopting TEQ means there is not a distinction between the work tasks and the nonwork tasks. The millennial does not think of communication with a client as work and making dinner reservations as leisure; it is all the same thing. Most important, by embracing TEQ, the millennial does not resent one part of his or her life interfering with the other because work and life have become one.

Apps and social media have facilitated the TEQ lifestyle with the millennial generation. At one time, apps and social media were seen as ways to play games on our phones, think Angry Birds, or an anonymous way to spy on people we dated in high school (we have all done it, or thought about doing it, via Facebook). Apps and social media have matured from being solely social game-playing mechanisms to both entertainment venues and business tools. Apps and social media successfully blend life and work via technology.

Currently, apps and social media's importance in the HR world is becoming more apparent. More than 90 percent of organizations use them for recruiting and greater than 70 percent have had favorable hiring outcomes through social media;[12] however, companies continue to place outdated rules or even social media bans on their employees. According to research firm Statista, one out of five companies block Facebook on company computers.[13] I have spoken with countless audience members who tell me their companies do everything from prohibiting cell phones in the office, to not allowing tablets in meetings, blocking all social media sites on company computers, or installing cameras in the break rooms with hopes of "catching" the younger generation using social media during office hours.[14]

These tactics are not only futile; they stifle the ability of millennials to take a TEQ approach while working for your organization. The rules also convey to the millennials that you do not trust them or value the skillset they bring to the workplace. More importantly, research is discovering people who use social media sites at work are more productive. A study by research firm Evolv shows that employees who access several social media sites retain more, get more done, and stay at their jobs longer.

What Can Organizations Do to Facilitate a TEQ Workplace for the Millennials?

First, get rid of bans on social media. Social media has become the smoke break of the 1970s and the water cooler chat of the 1980s. It is a mental reset for all employees, not just the millennials, to check in, connect and take a breather. Rename social media *social mental floss,* and you may feel better about the time people spend checking their Twitter feed.

This does not mean you can't have rules. A good social media policy is smart business. The rules need to make sense and apply equally to everyone.

For example, a social aid organization that caters to single parents has a ban on posting anything about the clients they serve. Sara McCarthy, a Gen X manager, says that, "Our clients talk about sensitive issues when they visit us. We do not want the clients to feel embarrassed or feel violated if they see pictures of themselves entering our offices or waiting in our lobby. We tell employees that because of these reasons, they cannot post anything about whom we serve at the office. We do, however, encourage employees to post how they feel about the work they do and post photos of themselves and coworkers."

Recently on Yelp, I read a scathing review of one of my favorite places to eat. The review did not criticize the food or service, but focused on one of the senior managers. The review took several mean jabs at the manager and seemed to be written by a disgruntled employee. Some organizations place social media rules for fear of negative comments that its employees may post. If that is the case, create an internal platform where your employees can vent without retribution. This not only allows people to "get it out of their system," it also gives management an opportunity to really "hear" what is plaguing their employees and take action.

Intel's social media policy is specific regarding what not to share, i.e., confidential information. But it also works from the assumption that their employees are mature individuals. Intel asks their employees to be "transparent, truthful, yourself, and up-to-date" when it comes to social media.[15] Intel further instructs their employees not to reveal confidential information, not to bash the competition, to admit when they make a mistake, and to correct their social media mistakes.[16]

Some organizations rely on employee discretion. Olivia Johnson (millennial) is a member service representative for a financial institution, stating that, "The company expects us to act responsibly when it comes to social media. They do not want us to get drunk and post pictures of ourselves everywhere. They expect us to use common sense."

Having a progressive social media policy is a critical factor for the millennials and their TEQ approach to work. Thirty-three percent of the millennials ranked "social media freedom" more important than salary.[17] As more and more of them join the workforce and assume management leadership positions, social media will become a larger part of the communication toolbox. Remember when it was all about the Three Rs—reading, writing, and arithmetic—of education? Now, we need to add social media to the equation. According to Future Workplace's Multiple Generations @ Work survey, by year 2020, 60 percent of the millennials believe that "social media literacy will be required of all employees."[18] It behooves all organizations to become more social with social media.

We Are People, Not Widgets. Reward Us as Individuals.

Carol Kauffman, VP and director of Development and Communications at Neighborhood Housing Services of Phoenix, and a baby boomer, recalls, "When I first began my career, my boss came back from vacation with a small gift for everyone on her team. She gave us all the same identical gift. At the time, I appreciated the thought and I know she did what she thought was fair. You can't do that with the younger generation today. Millennials expect to be motivated individually."

I had a "seasoned" audience member approach me after a presentation and tell me that motivation could be summed up in two words: "your paycheck."

It is tempting, especially following the great recession, to feel that people, especially younger people who lack years in the workforce, should express gratitude and be thankful for a paycheck and not concerned with other perks or rewards.

Millennials do not equate a large paycheck with job satisfaction. In a Brookings Institute study, over 60 percent of millennials claim that "they would rather make $40,000 a year at a job they love than $100,000 a year at a job they think is boring." According to the study, Understanding a Misunderstood Generation, the millennials feel pay is not as important as having more intangible benefits.[20] Almost half of the millennial generation would rather not have a job than work at a job they despised.[21]

A paycheck gets people in the front door. What keeps the millennials past their probationary period is more than just a salary. What motivates a person can be as unique as their fingerprint; however, millennials do have some common characteristics to consider when it comes to motivation.

Jaxson (millennial), an inside sales rep for a sports equipment manufacturer, states that, "I work here because I love sports. I work in a cube all day; I am not crazy about that aspect of my job, but the company rewards us and makes it easy to lead a healthy lifestyle, which is important to me. They have a health-wise program with a gym, locker room and showers, which are free to the employees. There is a chiropractor, coach, and a personal trainer. Every time you work out with your trainer, you get points and the points add up for additional vacation days. You can also earn extra points by getting a physical every year or visiting the dentist. One of the sales reps lost a considerable amount of weight using a Fitbit bracelet, so the company got everybody one. The company also encourages everyone to be smoke-free."

Not all organizations can afford as many perks as Jaxson's employer, but every company can create a bonus program that is not expensive, but still has value for millennials. For example, smaller companies allow employees to take a longer lunch if they are taking an exercise class during their lunch break.

Comp days are strong motivators for the millennial generation because it allows them flexibility. If your company can't afford to allow someone off for an entire day, give "slack hours"; time during the day when the employee can do whatever they want, such as update their social media, read, watch their dog online via their nanny-cam, or take a nap.

Other groups have small perks to create a happier work environment. Have a dry cleaning service pick up at the office, a massage therapist come in to give 15-minute chair massages, or (my favorite), bring in free food. These are the types of rewards that make the workplace a pleasant place to be.

The key to motivating the millennial generation is leaving the final look and feel of the reward in their hands. The more ownership they have over the reward, whether it's staying healthy, receiving flex-time or in-office perks, the greater the value, i.e., more motivating for the millennial generation.

They want their career and lifestyle to complement each other, they want TEQ, and they want to be motivated differently. They also want to make a difference and influence great change moving into the future. The good news for HR is that the millennials recognize that the greatest impact they can have on the world is through their employer. When we are willing to work differently with each other, only then will we take multiple generations and truly work successfully as one.

Notes

1. Dan Schawbel, "Millennials vs. Baby Boomers: Who Would You Rather Hire?" *Time,* March 29, 2012. http://business.time.com/2012/03/29/millennials-vs-baby-boomers-who-would-you-rather-hire/
2. Ibid.
3. Rob Asghar, "What Millennials Want in The Workplace (And Why You Should Start Giving It to Them)," *Forbes,* January 13, 2014. http://www.forbes.com/sites/robasghar/2014/01/13/what-millennials-want-in-the-workplace-and-why-you-should-start-giving-it-to-them/
4. Morley Winograd and Michael Hais, "How Millennials Could Upend Wall Street and Corporate America," May 2014. http://www.brookings.edu/~/media/research/files/papers/2014/05/millennials%20wall%20st/brookings_winogradv5.pdf

5. Executive Summary, "Millennials: Confident. Connected. Open to Change," Pew Research Center, February 24, 2010. http://www.pewsocialtrends.org/2010/02/24/millennials-confident-connected-open-to-change/
6. Ray Williams, "How the Millennial Generation Will Change the Workplace," *Psychology Today,* March 19, 2014. https://www.psychologytoday.com/blog/wired-success/201403/how-the-millennial-generation-will-change-the-workplace
7. Britt Hysen, "Millennials Making a Social Impact," *Huffington Post,* http://www.huffingtonpost.com/britt-hysen/millennials-making-a-soci_b_5851186.html.
8. http://www.businessdictionary.com/definition/work-life-balance.html.
9. Jesse Colombo, "Black Monday—the Stock Market Crash of 1987," The Bubble Bubble, August 3, 2012. http://www.thebubblebubble.com/1987-crash/
10. Millennials: Technology = Social Connection, Nielsen, February 26, 2014. http://www.nielsen.com/us/en/insights/news/2014/millennials-technology-social-connection.html
11. Carlos Dominguez, "The Smartphone Gets Which Side of the Bed?" Cisco, December 18, 2012. http://blogs.cisco.com/news/the-smartphone-gets-which-side-of-the-bed
12. Infographic, http://www.staff.com/blog/social-media-for-recruitment-infographic/, Staff.com, October 17, 2013.
13. Stephanie Vozza, "Why Banning Facebook in Your Workplace Is a Stupid Move," *Entrepreneur,* September 30, 2013. http://www.entrepreneur.com/article/228641
14. Francesca Fenzi, "Social Media: Not the Productivity Killer You Thought?" *Inc.,* http://www.inc.com/francesca-fenzi/social-media-not-the-productivity-killer-you-thought.html, April 4, 2013.
15. http://www.intel.com/content/www/us/en/legal/intel-social-media-guidelines.html.
16. Ibid.
17. Infographic, Staff.com, October 17, 2013. http://www.staff.com/blog/social-media-for-recruitment-infographic/
18. Jeanne Meister, "Want To Be A More Productive Employee? Get on Social Networks," *Forbes,* April 4, 2013. http://www.forbes.com/sites/jeannemeister/2013/04/18/want-to-be-a-more-productive-employee-get-on-social-networks/
19. Morley Winograd and Michael Hais, "How Millennials Could Upend Wall Street and Corporate America," Brookings Education, May 2014. http://www.brookings.edu/~/media/research/files/papers/2014/05/millennials%20wall%20st/brookings_winogradv5.pdf
20. Jessica Sier, "Generation Y thinks work-life balance more important than cash," *AFR,* November 14, 2014. http://www.afr.com/p/national/work_space/generation_thinks_work_life_balance_GwT1CXmsl7ZmY8T-Di31B8J

21. Emily Matchar, "How those spoiled millennials will make the workplace better for everyone," *Washington Post,* August 16, 2012. http://www.washingtonpost.com/opinions/how-those-spoiled-millennials-will-make-the-workplace-better-for-everyone/2012/08/16/814af692-d5d8-11e1-a0cc-8954acd5f90c_story.html

Understanding the Text

1. What is Technology Equilibrium? How does it allow the millennial generation to work differently from previous generations?
2. According to Johnson, what happens when companies prohibit technology usage at work? How do millennials feel? Are you surprised? Why or why not?
3. What are some key differences in the workplace, according to Johnson, between millennials and baby boomers?

Reflection and Response

4. Johnson explains that millennials often do not see a separation between their work and personal life because technology makes it too easy to toggle back and forth between the two. What are some of the positive and negative effects of this millennial characteristic? Does Johnson make this seem more like a pro or a con? Provide evidence from the reading.
5. Based on Johnson's descriptions of *how* millennials work, do millennials seem more efficient at work than previous generations? Why or why not?
6. Find the article "Generation Z: What They Want in the Workplace" by Erika Morphy on the website cmswire.com. Compare Morphy's article about Generation Z to the claims Johnson makes about millennials in this article. As Generation Z continues to grow, what changes do you foresee happening in the workplace?

Making Connections

7. Johnson writes that "millennials are more ethnically diverse than past generations, more racially accepting, more likely to have a tattoo, more open to immigration than older generations" (par. 3). How do you think technology has played a role in helping the millennial generation develop more acceptance and inclusivity?
8. This article argues that millennials aren't simply after a big paycheck or a corner office; Johnson says that millennials often want to work for companies that are trying to improve the world we live in. Think back to other readings from this chapter, particularly looking at what Amazon is doing and the idea of a peer-to-peer economy. Do those readings, by Fickenscher (p. 198) and Britton (p. 217) respectively, support Johnson's claims? Why or why not?

9. Johnson portrays the millennial generation as a group of workers with great potential, who simply need the right motivation to work to help companies succeed; however, other scholars, like psychologist Jean Twenge, do not always discuss millennials in such a positive tone. Research how technology is affecting the millennial generation at work, and write an essay arguing whether or not technology is harming them more than helping them in the workplace.

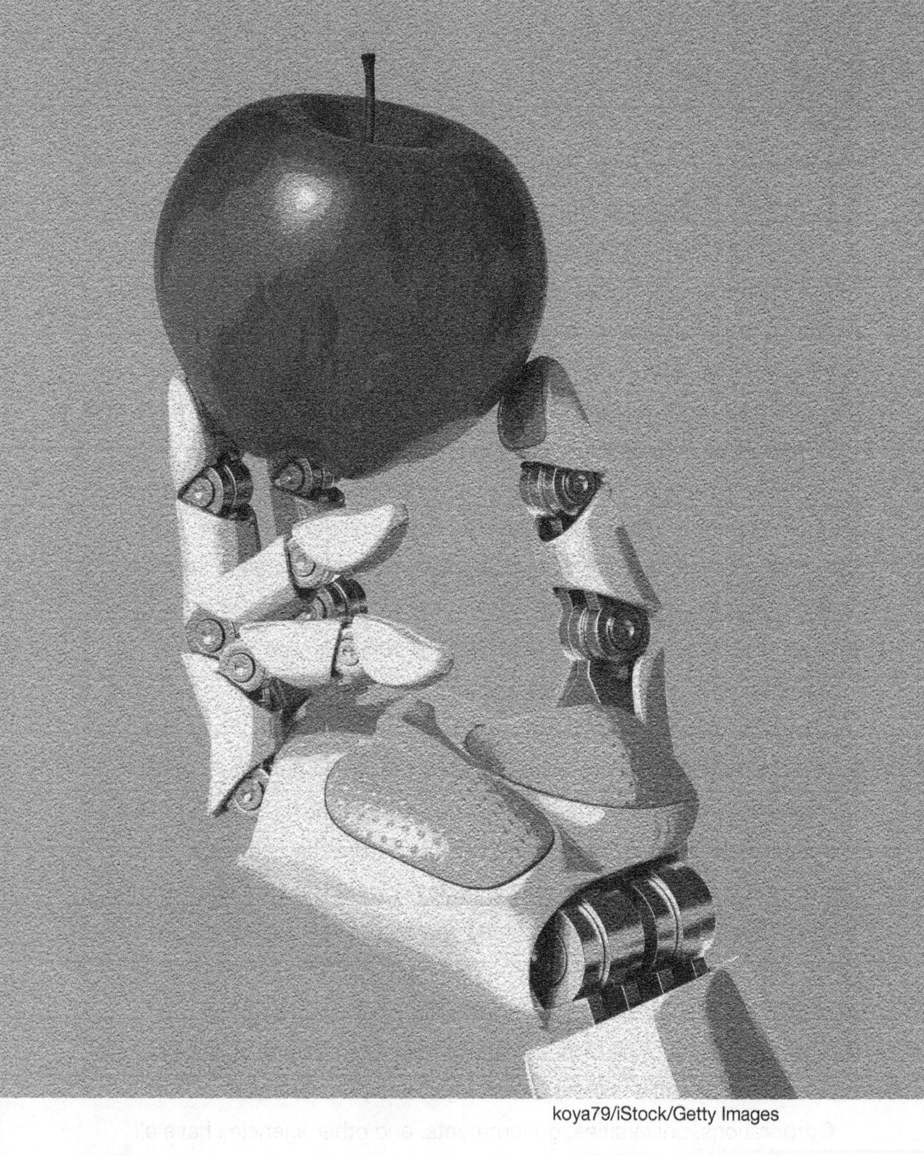

koya79/iStock/Getty Images

5 How Is the Internet Defining What Matters to Society?

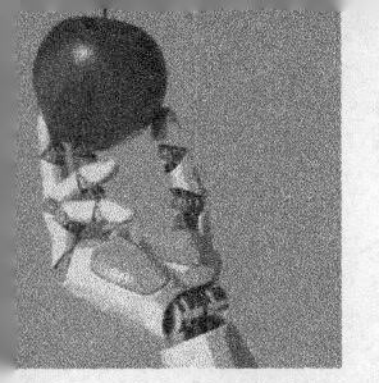

The definition of community typically includes a group of people with shared interests, and when we think about the various communities that exist on or because of the internet, we can often see ourselves fitting easily within at least one. Take for instance the communities developed on social media platforms like Instagram where users follow, tag, and share with one another regularly. Although these followers may be perfect strangers outside of Instagram, users have created rich and diverse communities *within* the platform.

Online communities such as this one have impacted society in many ways. For example, users can make new friends and connect with old ones on Facebook, discover and share news on social media outlets like Twitter — as we see discussed in "How Americans Get Their News" — or build and conduct business on entrepreneur-friendly online commerce sites like Etsy. Such communities also offer us ways to engage in activism. For example, the readings "Small Change: Why the Revolution Will Not Be Tweeted" and "Sorry, Malcolm Gladwell, the Revolution May Well Be Tweeted" help us better understand campaigns that are born online through hashtags and then travel offline in the forms of rallies and protests. Even people who do not feel particularly tech-savvy would find it hard to argue that their lives are not touched in some way by the internet, as we see so clearly explained in "When Your Smartphone Is Too Smart for Your Own Good: How Social Media Alters Human Relationships" and "How Has Technology Changed the Concept of Community?," both of which examine the idea that our entire society is now affected by the internet.

Corporations, universities, governments, and other agencies have all made use of these communities, too, with their own agendas and goals. Local sheriffs' departments provide updates through apps like Nextdoor and Patch, news organizations post breaking news stories to Twitter faster than their cable or paper equivalents can get the word out, foreign governments can control state-sanctioned messages via Facebook, universities such as Harvard track disease outbreaks through social media,

photo: koya79/iStock/Getty Images

and new currencies like Bitcoin exist solely online. But how do these actions that take place in virtual communities affect individuals on and offline? Why might that matter?

In "Net Neutrality and the Fight for Social Justice," we are reminded that the internet is often portrayed as an equalizer, a piece of technology that is supposed to allow all people to exercise their voices, and, yes, sometimes that's true, even for those who may have their voices suppressed offline; however, it is easy to challenge this idea when we examine who really has affordable and equal access to the internet and who controls that access. And even if there was a Utopian Internet-community where everyone had access, the article "Can Artificial Intelligence Help Solve the Internet's Misinformation Problem?" reminds us that we must be careful what we believe as we encounter news online, especially when a lie can travel so far so quickly in an online environment. As you read this chapter, think about how the internet has benefited community and relationship building while also thinking about the ways it excludes people, people who may be getting excluded twice: in reality and in virtual reality.

How Americans Get Their News

The Media Insight Project

The Media Insight Project was created through a partnership between the American Press Institute and the AP-NORC Center for Public Affairs Research based at the University of Chicago. The project's goal is to learn more about news consumers and use that knowledge to better inform news organizations.

The following study reports on how digital news and media are affecting the ways society today receives its news. The article explains how the availability of technology, including the prevalence of individuals who own multiple devices, is expanding people's access to news. While traditional news sources like newspapers and magazines have seen their physical subscriptions dwindling, this does not mean consumers are any less interested in current events — it simply means that they are moving online to find them. But are these additional options improving the way we ingest the news? As you read, think about the effects, both positive and negative, technology is having on news consumption today.

Americans are Accessing the News throughout the Day and across Devices

For many Americans, keeping up with the news is an activity that occurs throughout the day and across different formats, devices, and technologies. When asked when they prefer to watch, read, or hear news, a plurality (33 percent) report following the news all throughout the day. A smaller but sizable number of Americans continue to prefer to follow the news in the morning (24 percent) and in the evening (26 percent), while still lesser numbers say they most often get news in the afternoon (4 percent) and right before bed (9 percent).

Is there a particular time that people go for more in-depth news, beyond the headlines? Overall, four in ten Americans report that they delved deeper into a particular news subject beyond the headlines in the last week. When they did, that in-depth reading, watching, or listening followed a similar pattern to news consumption generally, with a plurality (34 percent) saying there is no particular time they prefer to read in-depth news. That finding challenges the notion that while Americans may get headlines continuously, they reserve the evening for learning more.

In addition, a slightly larger number, 49 percent of adults, said they delved deeper to learn more about the last breaking news story they paid attention to, though time of day was not probed for this, given that news may break at any time.

Americans follow the news on a wide variety of devices, including through television, radio, print versions of newspapers and magazines, computers, cell phones, tablets, e-readers, and devices such as an Xbox or Playstation that link the internet to a television. Americans on average reported that, during the past week, they followed the news using four different devices or technologies. The most frequently utilized devices include television (87 percent), laptops/computers (69 percent), radio (65 percent), and print newspapers or magazines (61 percent).

Among all adult Americans, 56 percent reported using a cell phone and 29 percent reported using a tablet to access news in the last week. (That represents 78 percent of the 69 percent of Americans who own a smartphone, and 73 percent of the 39 percent of Americans who own or use a tablet device.) 5

Only 10 percent of Americans reported using an e-reader to get their news in the last week and 11 percent reported using a Smart TV to follow the news in the last week.

Traditional media remain important even for those Americans with the most gadgets. People who own and use more devices are no more or less likely to use print publications, television, or radio to access the news. For example, 62 percent of people who use only one internet enabled device say they used the print version of a publication to get news in the last week, as do 60 percent of those who use three or more mobile devices.

However, as the number of devices a person owns increases, they are more likely to report that they enjoy keeping up with the news and are more likely to say that it's easier to keep up with the news today than it was five years ago.

In addition to asking Americans about all the devices they use to get the news, the survey asked what device or technology people prefer most. Television was most popular (24 percent), followed by desktop or laptop computer (12 percent), cell phone (12 percent), and tablet computer (4 percent).

The largest group, 45 percent of Americans, indicate that they have no preference in the device or technology they use to follow the news. This suggests that many Americans prefer to receive news across devices, using whatever device or technology is most convenient when they want to follow the news. 10

People Use a Number of Different Tools to Find Out about the News but Trust Some More Than Others

Regardless of the device used, people find or discover news through different means—from old-fashioned word-of-mouth to electronic alerts and social media. The survey reveals that most Americans are discovering news in more than one way. More than half of all Americans report using between three and five methods of discovery to find out about the news (the average across all respondents was between three and four different methods in the last week [mean = 3.8]).

The most popular way that Americans report finding their news is directly from a news organization, such as a newspaper, TV newscast, website, or newswire (88 percent). People continue to discover news through traditional word-of-mouth (65 percent) either in person or over the phone, and do so at higher rates than more modern methods of sharing like email, text message, or other ways online (46 percent), or social media (44 percent). And roughly half of Americans said they got news in the last week from search engines and online news aggregators (51 percent for each).

But while Americans are discovering news through a variety of means, they are discriminating in how much trust they have in the information they get from each method.

China's Xinhua News agency debuted an AI news anchor in 2018. His speech is based on real-life Xinhua news anchor, Zhang Zhao.
Kyodo/AP Images

Americans trust the news they hear directly from news-gathering organizations, with 43 percent of people saying they trust the information acquired this way either very much or completely, 44 percent saying they trust it moderately, and 13 percent saying they only trust it slightly or not at all.

Between discovering news through aggregators or search engines, Americans who use them say they have more faith in search engines. In all, 32 percent who get news via search engines say they trust the information they provide either very much or completely compared to just 24 percent who use news aggregators.

And generally, levels of trust in the news discovered through sharing—either verbally or electronically—are low. Just 27 percent say they trust a good deal (very much or completely) the information they receive through electronic sharing with friends, and 21 percent have similar levels of trust for information they receive by word-of-mouth.

Similarly, while social media is becoming an important means of discovering news, even those who use it bring some skepticism to it. Only 15 percent of adults who get news through social media say they have high levels of trust in information they get from that means of discovery. Social media and word-of-mouth are the least-trusted means of discovering the news, with 37 percent of those who got news this way in the last week mistrusting or trusting only slightly social media and 33 percent mistrusting word-of-mouth.

Less than one-third (31 percent) of Americans report that they discovered the news in the last week through electronic news alerts or subscriptions they've signed up for (more Americans, 47 percent, report that they *ever* receive news alerts through text, email, or apps), but they trust this information at higher rates (50 percent) than any other discovery method. Most people who get news from electronic news alerts (92 percent) also report getting information directly from news organizations that report the news.

The survey data also show the powerful connection between the growth in mobile internet technology and social media. Those adults with a cell phone that connects to the internet are much more likely than those without one to find news through social media (56 percent vs. 22 percent).

Similarly, adults with smartphones are also much more likely than non-mobile users to get news through search engines (61 percent vs. 31 percent); online news aggregators (61 percent vs. 33 percent); sharing with friends via email, text, or other online methods (54 percent vs. 29 percent); and electronic news alerts (38 percent vs. 19 percent). Similar patterns emerged for owners or users of tablets. Tablet owners are also more likely to report discovering news through social media, search engines, online news aggregators, electronic communications with friends, and news alerts than people who don't own tablets.

For the most part, mobile users are not more likely than non-users to trust electronic means of discovering the news. Levels of trust in social media, search engines, electronic communications with friends, and news alerts are similar between users and non-users. The one exception is that smartphone owners are more likely than non-owners to say they trust news from online aggregators (27 percent vs. 17 percent).

While people use different media for news each week, and only half express a preference for a device, most Americans do care about the means of discovery, or where they first hear the news. More than six in ten Americans (61 percent) say they prefer getting news directly from a news organization compared with any other way.

That preference dwarfs all other discovery methods. Every other means of discovering the news was preferred by fewer than one in ten adults, with search engines (7 percent), online publishers that mostly combine news from other sources (7 percent), and social media (4 percent), in order of preference.

Similarly, while Americans cite electronic news alerts as a highly trusted method of discovering the news, only 3 percent volunteer news alerts as their most preferred way to discover the news. (Notably, nearly everyone, 92 percent, who gets news from electronic news alerts also reports getting news directly from news organizations.) Word-of-mouth in person or over the phone (2 percent) and sharing news with friends through email, text message, or other ways online (1 percent) are cited by very few Americans as their most preferred way to get their news.

While People are Getting News Across a Number of Devices and Tools, They are Discriminating in Terms of the Source of the News and Whose News They Trust

Beyond the way in which Americans discover the news and the devices they use to access it, people also have choices in the source of their news—the type of news organization doing the reporting that they turn to. And, when it comes to who does the reporting, Americans don't tend to rely on a single source. The survey finds that the average American recalled getting their news from between four and five of eight different types of news sources in the last week (mean = 4.56). And, similar to the way they discover news, news consumers discriminate between reporting sources in terms of their level of trust.

Television news organizations are the most popular news source for Americans. Whether from the TV broadcast or the station's website,

93 percent of Americans say they used some kind a TV news operation as a source of news in the last week.

Among the different types of TV news, more Americans (82 percent) turn to their local TV news stations either through the TV broadcast or online than another other type. Sizeable majorities also cite the three national network broadcast news operations (73 percent) in their various forms and 24-hour cable news channels (62 percent)—such as Fox News, CNN, or MSNBC—as sources of news, either on television or digitally.

A majority of Americans also cite newspaper content in its various forms (66 percent) and radio-based news reporting sources (56 percent) as news sources they accessed in the past week.

Slightly less than half of Americans (47 percent) say they used online-only reporting sources such as *Yahoo! News, BuzzFeed, The Huffington Post,* or other blogs in the last week. Thirty-seven percent report using magazines—print or online—as a source of news in the last week. And, a third of Americans say they now get news from wire services such as The Associated Press (AP) or Reuters, something that was not easy to do directly before the internet.

Yet people have varying levels of trust in these different sources, and, for the most part, only about half or less of Americans say they very much or completely trust any of these sources.

Overall, Americans report that they trust the information they get from local TV news stations to a greater degree than any other source of news, with 52 percent who seek out local TV news saying that they trust the information very much or completely. At similar levels, 51 percent of those who use the newswires say they trust them, 48 percent trust radio news, and 47 percent trust newspapers and the three broadcast networks, NBC, ABC, and CBS. Forty-four percent of those who use cable news say they have a high level of trust in it. Users of magazines (print or online) as a source of news report slightly more modest levels of trust (40 percent completely or very much).

"Overall, Americans report that they trust the information they get from local TV news stations to a greater degree than any other source of news."

Online-only sources of the news such as *Yahoo! News, BuzzFeed,* or *The Huffington Post,* and blogs garner lower levels of trust. One in four users of these news sources say they trust them completely or very much, while one in five users say they trust them only slightly or not at all.

Contrary to the idea that people now tend to trust news sources that share their point of view, taken together the findings suggest that rates of trust are highest for news operations that have less editorial opinion built into their model, such as local television news and wire services.

Again the findings suggest that news consumers with a plethora of choices are discriminating—utilizing sources that fit their habits. For instance, Americans who report that they watch, read, or hear the news at least once a day are more likely than others to cite a 24-hour TV news channel as a reporting source they use (67 percent vs. 50 percent). Daily news consumers are also more likely than others to cite getting news from radio news organizations (60 percent vs. 47 percent) and newswires such as AP or Reuters (37 percent vs. 21 percent).

Further, cable news consumers are also more likely to say they get news "throughout the day" rather than at defined times of day. Nearly three-fourths (73 percent) of cable news consumers say they prefer to watch, read, or hear the news throughout the day, significantly higher than those who say "in the morning" (58 percent), "in the evening" (56 percent), or "the last thing at night" (51 percent). Similarly, those who get news from newswires such as AP or Reuters tend to be continuous news consumers (42 percent throughout the day vs. 32 percent in the morning, 25 percent in the evening, and 26 percent last thing at night). 35

Mobile technology, similarly, correlates with heavier use of non-traditional sources. People with a smartphone are much more likely than those who do not have smartphones to say they get news from online-only sources like *Yahoo! News, BuzzFeed, The Huffington Post*, or other blogs (58 percent vs. 26 percent). Mobile news consumers are also more likely than others to say they get news directly from newswires such as the AP or Reuters (36 percent vs. 27 percent).

Understanding the Text

1. According to the reading, which type of news source do Americans trust most? Which type is used the most?
2. Does the study reveal that having access to a variety of different news sources makes users more or less discriminating? Why do you think that is?

Reflection and Response

3. The article contends that many Americans prefer gathering news via multiple technological devices. Identify both negatives and positives to this approach.
4. Review the statistics in the reading and identify which avenue people *least* prefer for news. Why would that method of gathering news be least preferred? Use the reading and personal experience to support your response.

5. Only 2 percent of Americans cite word-of-mouth as their preferred way to gather news according to the reading (par. 24). Why do you think this is so? Could this be problematic? Why or why not?

Making Connections

6. Malcolm Gladwell's "Small Change: Why the Revolution Will Not Be Tweeted," is the next reading in the chapter, and in it he argues that social media engenders lazy activism. The Media Insight Project's study shows an increasing number of people receive their news from social media. If both Gladwell and the results of this survey are to be believed, how do you think this shift will affect American engagement in activism? Support your response with evidence from both sources and scholarly research.
7. Locate and read the article "Is Google Making Us Stupid?" by Nicholas Carr, published by *The Atlantic*. Carr's piece says that people read more today than ever before, but he argues that it's not quality reading. Carr says we skim, click, and move on, often without even reading some articles at all. Write an essay, analyzing and synthesizing the information from his article and the results of this study, developing conclusions about how technology is both helping and hurting our ability to read and receive news.

Small Change: Why the Revolution Will Not Be Tweeted

Malcolm Gladwell

Malcolm Gladwell is an author, journalist, and podcast host. A staff writer for the *New Yorker* since 1996, Gladwell often writes about topics related to the social sciences, and his various books can be interpreted through psychological, sociological, and/or social psychological lenses, helping his ideas span the disciplines. Gladwell has written five best-selling books, including *The Tipping Point* (2000) and *Outliers* (2008), and was named one of the 100 most influential people by *Time* magazine in 2005.

In the following article, Gladwell argues that today's revolutions will not run as smoothly as revolutions prior to social media. Gladwell believes social media lends itself to "lazy activism," and fails to help us build substantial ties with each other. As you read, think about whether or not you agree with Gladwell's ideas and how your own personal observations help support Gladwell or challenge his claims.

At four-thirty in the afternoon on Monday, February 1, 1960, four college students sat down at the lunch counter at the Woolworth's in downtown Greensboro, North Carolina. They were freshmen at North Carolina A. & T., a black college a mile or so away.

"I'd like a cup of coffee, please," one of the four, Ezell Blair, said to the waitress.

"We don't serve Negroes here," she replied.

The Woolworth's lunch counter was a long L-shaped bar that could seat sixty-six people, with a standup snack bar at one end. The seats were for whites. The snack bar was for blacks. Another employee, a black woman who worked at the steam table, approached the students and tried to warn them away. "You're acting stupid, ignorant!" she said. They didn't move. Around five-thirty, the front doors to the store were locked. The four still didn't move. Finally, they left by a side door. Outside, a small crowd had gathered, including a photographer from the Greensboro *Record*. "I'll be back tomorrow with A. & T. College," one of the students said.

5 By next morning, the protest had grown to twenty-seven men and four women, most from the same dormitory as the original four. The men were dressed in suits and ties. The students had brought their schoolwork, and studied as they sat at the counter. On Wednesday, students from Greensboro's "Negro" secondary school, Dudley High, joined in,

and the number of protesters swelled to eighty. By Thursday, the protesters numbered three hundred, including three white women, from the Greensboro campus of the University of North Carolina. By Saturday, the sit-in had reached six hundred. People spilled out onto the street. White teenagers waved Confederate flags. Someone threw a firecracker. At noon, the A. & T. football team arrived. "Here comes the wrecking crew," one of the white students shouted.[1]

By the following Monday, sit-ins had spread to Winston-Salem, twenty-five miles away, and Durham, fifty miles away. The day after that, students at Fayetteville State Teachers College and at Johnson C. Smith College, in Charlotte, joined in, followed on Wednesday by students at St. Augustine's College and Shaw University, in Raleigh. On Thursday and Friday, the protest crossed state lines, surfacing in Hampton and Portsmouth, Virginia, in Rock Hill, South Carolina, and in Chattanooga, Tennessee. By the end of the month, there were sit-ins throughout the South, as far west as Texas. "I asked every student I met what the first day of the sitdowns had been like on his campus," the political theorist Michael Walzer wrote in *Dissent*. "The answer was always the same: 'It was like a fever. Everyone wanted to go.'" Some seventy thousand students eventually took part. Thousands were arrested and untold thousands more radicalized. These events in the early sixties became a civil-rights war that engulfed the South for the rest of the decade—and it happened without e-mail, texting, Facebook, or Twitter.

> "The new tools of social media have reinvented social activism."

The world, we are told, is in the midst of a revolution. The new tools of social media have reinvented social activism. With Facebook and Twitter and the like, the traditional relationship between political authority and popular will has been upended,° making it easier for the powerless to collaborate, coördinate, and give voice to their concerns. When ten thousand protesters took to the streets in Moldova in the spring of 2009 to protest against their country's Communist government, the action was dubbed the Twitter Revolution, because of the means by which the demonstrators had been brought together. A few months after that, when student protests rocked Tehran, the State Department took the unusual step of asking Twitter to suspend scheduled maintenance of its Web site, because the Administration didn't want such a critical organizing tool out of service at the height of the demonstrations. "Without Twitter the people of Iran would not have felt empowered and confident to stand up for freedom

Upended: turning something upside down or on its head.

and democracy," Mark Pfeifle, a former national-security adviser, later wrote, calling for Twitter to be nominated for the Nobel Peace Prize. Where activists were once defined by their causes, they are now defined by their tools. Facebook warriors go online to push for change. "You are the best hope for us all," James K. Glassman, a former senior State Department official, told a crowd of cyber activists at a recent conference sponsored by Facebook, AT&T, Howcast, MTV, and Google. Sites like Facebook, Glassman said, "give the U.S. a significant competitive advantage over terrorists. Some time ago, I said that Al Qaeda was 'eating our lunch on the Internet.' That is no longer the case. Al Qaeda is stuck in Web 1.0. The Internet is now about interactivity and conversation."

These are strong, and puzzling, claims. Why does it matter who is eating whose lunch on the Internet? Are people who log on to their Facebook page really the best hope for us all? As for Moldova's so-called Twitter Revolution, Evgeny Morozov, a scholar at Stanford who has been the most persistent of digital evangelism's critics, points out that Twitter had scant internal significance in Moldova, a country where very few Twitter accounts exist. Nor does it seem to have been a revolution, not least because the protests—as Anne Applebaum suggested in *The Washington Post*—may well have been a bit of stagecraft cooked up by the government. (In a country paranoid about Romanian revanchism°, the protesters flew a Romanian flag over the Parliament building.) In the Iranian case, meanwhile, the people tweeting about the demonstrations were almost all in the West. "It is time to get Twitter's role in the events in Iran right," Golnaz Esfandiari wrote, this past summer, in *Foreign Policy*. "Simply put: There was no Twitter Revolution inside Iran." The cadre of prominent bloggers, like Andrew Sullivan, who championed the role of social media in Iran, Esfandiari continued, misunderstood the situation. "Western journalists who couldn't reach—or didn't bother reaching?—people on the ground in Iran simply scrolled through the English-language tweets post with tag #iranelection," she wrote. "Through it all, no one seemed to wonder why people trying to coordinate protests in Iran would be writing in any language other than Farsi."

Some of this grandiosity is to be expected. Innovators tend to be solipsists.° They often want to cram every stray fact and experience into their new model. As the historian Robert Darnton has written, "The marvels of communication technology in the present have produced a false consciousness about the past—even a sense that communication has no history, or had nothing of importance to consider before the days

Revanchism: political retaliation.
Solipsists: people who believe only the self exists.

of television and the Internet." But there is something else at work here, in the outsized enthusiasm for social media. Fifty years after one of the most extraordinary episodes of social upheaval in American history, we seem to have forgotten what activism is.

Greensboro in the early nineteen-sixties was the kind of place where racial insubordination was routinely met with violence. The four students who first sat down at the lunch counter were terrified. "I suppose if anyone had come up behind me and yelled 'Boo,' I think I would have fallen off my seat," one of them said later. On the first day, the store manager notified the police chief, who immediately sent two officers to the store. On the third day, a gang of white toughs showed up at the lunch counter and stood ostentatiously behind the protesters, ominously muttering epithets such as "burr-head nigger." A local Ku Klux Klan leader made an appearance. On Saturday, as tensions grew, someone called in a bomb threat, and the entire store had to be evacuated. 10

The dangers were even clearer in the Mississippi Freedom Summer Project of 1964, another of the sentinel campaigns of the civil-rights movement. The Student Nonviolent Coordinating Committee recruited hundreds of Northern, largely white unpaid volunteers to run Freedom Schools, register black voters, and raise civil-rights awareness in the Deep South. "No one should go *anywhere* alone, but certainly not in an automobile and certainly not at night," they were instructed. Within days of arriving in Mississippi, three volunteers—Michael Schwerner, James Chaney, and Andrew Goodman—were kidnapped and killed, and, during the rest of the summer, thirty-seven black churches were set on fire and dozens of safe houses were bombed; volunteers were beaten, shot at, arrested, and trailed by pickup trucks full of armed men. A quarter of those in the program dropped out. Activism that challenges the status quo—that attacks deeply rooted problems—is not for the faint of heart.

What makes people capable of this kind of activism? The Stanford sociologist Doug McAdam compared the Freedom Summer dropouts with the participants who stayed, and discovered that the key difference wasn't, as might be expected, ideological fervor. "*All* of the applicants—participants and withdrawals alike—emerge as highly committed, articulate supporters of the goals and values of the summer program," he concluded. What mattered more was an applicant's degree of personal connection to the civil-rights movement. All the volunteers were required to provide a list of personal contacts—the people they wanted kept apprised of their activities—and participants were far more likely than dropouts to have close friends who were also going to Mississippi. High-risk activism, McAdam concluded, is a "strong-tie" phenomenon.

This pattern shows up again and again. One study of the Red Brigades, the Italian terrorist group of the nineteen-seventies, found that seventy per cent of recruits had at least one good friend already in the organization. The same is true of the men who joined the mujahideen in Afghanistan. Even revolutionary actions that look spontaneous, like the demonstrations in East Germany that led to the fall of the Berlin Wall, are, at core, strong-tie phenomena. The opposition movement in East Germany consisted of several hundred groups, each with roughly a dozen members. Each group was in limited contact with the others: at the time, only thirteen per cent of East Germans even had a phone. All they knew was that on Monday nights, outside St. Nicholas Church in downtown Leipzig, people gathered to voice their anger at the state. And the primary determinant of who showed up was "critical friends"—the more friends you had who were critical of the regime the more likely you were to join the protest.

So one crucial fact about the four freshmen at the Greensboro lunch counter—David Richmond, Franklin McCain, Ezell Blair, and Joseph McNeil—was their relationship with one another. McNeil was a roommate of Blair's in A. & T.'s Scott Hall dormitory. Richmond roomed with McCain one floor up, and Blair, Richmond, and McCain had all gone to Dudley High School. The four would smuggle beer into the dorm and talk late into the night in Blair and McNeil's room. They would all have remembered the murder of Emmett Till in 1955, the Montgomery bus boycott that same year, and the showdown in Little Rock in 1957. It was McNeil who brought up the idea of a sit-in at Woolworth's. They'd discussed it for nearly a month. Then McNeil came into the dorm room and asked the others if they were ready. There was a pause, and McCain said, in a way that works only with people who talk late into the night with one another, "Are you guys chicken or not?" Ezell Blair worked up the courage the next day to ask for a cup of coffee because he was flanked by his roommate and two good friends from high school.

15 The kind of activism associated with social media isn't like this at all. The platforms of social media are built around weak ties. Twitter is a way of following (or being followed by) people you may never have met. Facebook is a tool for efficiently managing your acquaintances, for keeping up with the people you would not otherwise be able to stay in touch with. That's why you can have a thousand "friends" on Facebook, as you never could in real life.

This is in many ways a wonderful thing. There is strength in weak ties, as the sociologist Mark Granovetter has observed. Our acquaintances—not our friends—are our greatest source of new ideas and information. The Internet lets us exploit the power of these kinds of distant connections with marvellous efficiency. It's terrific at the diffusion of innovation, interdisciplinary collaboration, seamlessly matching up buyers and sellers, and the logistical functions of the dating world. But weak ties seldom lead to high-risk activism.

In a new book called "The Dragonfly Effect: Quick, Effective, and Powerful Ways to Use Social Media to Drive Social Change," the business consultant Andy Smith and the Stanford Business School professor Jennifer Aaker tell the story of Sameer Bhatia, a young Silicon Valley entrepreneur who came down with acute myelogenous leukemia. It's a perfect illustration of social media's strengths. Bhatia needed a bone-marrow transplant, but he could not find a match among his relatives and friends. The odds were best with a donor of his ethnicity, and there were few South Asians in the national bone-marrow database. So Bhatia's business partner sent out an e-mail explaining Bhatia's plight to more than four hundred of their acquaintances, who forwarded the e-mail to their personal contacts; Facebook pages and YouTube videos were devoted to the Help Sameer campaign. Eventually, nearly twenty-five thousand new people were registered in the bone-marrow database, and Bhatia found a match.

But how did the campaign get so many people to sign up? By not asking too much of them. That's the only way you can get someone you don't really know to do something on your behalf. You can get thousands of people to sign up for a donor registry, because doing so is pretty easy. You have to send in a cheek swab and—in the highly unlikely event that your bone marrow is a good match for someone in need—spend a few hours at the hospital. Donating bone marrow isn't a trivial matter. But it doesn't involve financial or personal risk; it doesn't mean spending a summer being chased by armed men in pickup trucks. It doesn't require that you confront socially entrenched norms and practices. In fact, it's the kind of commitment that will bring only social acknowledgment and praise.

The evangelists of social media don't understand this distinction; they seem to believe that a Facebook friend is the same as a real friend and that signing up for a donor registry in Silicon Valley today is activism in the same sense as sitting at a segregated lunch counter in Greensboro in 1960. "Social networks are particularly effective at increasing motivation," Aaker and Smith write. But that's not true. Social networks are effective at increasing *participation*—by lessening the level of motivation that participation requires. The Facebook page of the Save Darfur Coalition has 1,282,339 members, who have donated an average of nine cents apiece. The next biggest Darfur charity on Facebook has 22,073 members, who have donated an average of thirty-five cents. Help Save Darfur has 2,797 members, who have given, on average, fifteen cents. A spokesperson for the Save Darfur Coalition told *Newsweek,* "We wouldn't necessarily gauge someone's value to the advocacy movement based on what they've given. This is a powerful mechanism to engage this critical population. They inform their community, attend events, volunteer. It's not something you can measure by looking at a ledger." In other words, Facebook activism succeeds not by motivating people to make a real sacrifice but by motivating them to do the things that people do when they are not motivated enough to make a real sacrifice. We are a long way from the lunch counters of Greensboro.

The students who joined the sit-ins across the South during the winter of 1960 described the movement as a "fever." But the civil-rights movement was more like a military campaign than like a contagion. In the late nineteen-fifties, there had been sixteen sit-ins in various cities throughout the South, fifteen of which were formally organized by civil-rights organizations like the N.A.A.C.P. and CORE. Possible locations for activism were scouted. Plans were drawn up. Movement activists held training sessions and retreats for would-be protesters. The Greensboro Four were a product of this groundwork: all were members of the N.A.A.C.P. Youth Council. They had close ties with the head of the local N.A.A.C.P. chapter. They had been briefed on the earlier wave of sit-ins in Durham, and had been part of a series of movement meetings in activist churches. When the sit-in movement spread from Greensboro throughout the South, it did not spread indiscriminately. It spread to those cities which had preëxisting "movement centers"—a core of dedicated and trained activists ready to turn the "fever" into action. 20

The civil-rights movement was high-risk activism. It was also, crucially, strategic activism: a challenge to the establishment mounted with precision and discipline. The N.A.A.C.P. was a centralized organization, run from New York according to highly formalized operating procedures. At the Southern Christian Leadership Conference, Martin Luther King, Jr., was the unquestioned authority. At the center of the movement was the black church, which had, as Aldon D. Morris points out in his superb 1984 study, "The Origins of the Civil Rights Movement," a carefully demarcated division of labor, with various standing committees and disciplined groups. "Each group was task-oriented and coordinated its activities through authority structures," Morris writes. "Individuals were held accountable for their assigned duties, and important conflicts were resolved by the minister, who usually exercised ultimate authority over the congregation."

This is the second crucial distinction between traditional activism and its online variant: social media are not about this kind of hierarchical organization. Facebook and the like are tools for building *networks*, which are the opposite, in structure and character, of hierarchies. Unlike hierarchies, with their rules and procedures, networks aren't controlled by a single central authority. Decisions are made through consensus, and the ties that bind people to the group are loose.

This structure makes networks enormously resilient and adaptable in low-risk situations. Wikipedia is a perfect example. It doesn't have an editor, sitting in New York, who directs and corrects each entry. The effort of putting together each entry is self-organized. If every entry in Wikipedia were to be erased tomorrow, the content would swiftly be restored, because that's what happens when a network of thousands spontaneously devote their time to a task.

There are many things, though, that networks don't do well. Car companies sensibly use a network to organize their hundreds of suppliers, but not to design their cars. No one believes that the articulation of a coherent design philosophy is best handled by a sprawling, leaderless organizational system. Because networks don't have a centralized leadership structure and clear lines of authority, they have real difficulty reaching consensus and setting goals. They can't think strategically; they are chronically prone to conflict and error. How do you make difficult choices about tactics or strategy or philosophical direction when everyone has an equal say?

25 The Palestine Liberation Organization originated as a network, and the international-relations scholars Mette Eilstrup-Sangiovanni and Calvert Jones argue in a recent essay in *International Security* that this is why it ran into such trouble as it grew: "Structural features typical of networks—the absence of central authority, the unchecked autonomy of rival groups, and the inability to arbitrate quarrels through formal mechanisms—made the P.L.O. excessively vulnerable to outside manipulation and internal strife."

In Germany in the nineteen-seventies, they go on, "the far more unified and successful left-wing terrorists tended to organize hierarchically, with professional management and clear divisions of labor. They were concentrated geographically in universities, where they could establish central leadership, trust, and camaraderie through regular, face-to-face meetings." They seldom betrayed their comrades in arms during police interrogations. Their counterparts on the right were organized as decentralized networks, and had no such discipline. These groups were regularly infiltrated, and members, once arrested, easily gave up their comrades. Similarly, Al Qaeda was most dangerous when it was a unified hierarchy. Now that it has dissipated into a network, it has proved far less effective.

The drawbacks of networks scarcely matter if the network isn't interested in systemic change—if it just wants to frighten or humiliate or make a splash—or if it doesn't need to think strategically. But if you're taking on a powerful and organized establishment you have to be a hierarchy. The Montgomery bus boycott required the participation of tens of thousands of people who depended on public transit to get to and from work each day. It lasted a *year*. In order to persuade those people to stay true to the cause, the boycott's organizers tasked each local black church with maintaining morale, and put together a free alternative private carpool service, with forty-eight dispatchers and forty-two pickup stations. Even the White Citizens Council, King later said, conceded that the carpool system moved with "military precision." By the time King came to Birmingham, for the climactic showdown with Police Commissioner Eugene (Bull) Connor, he had a budget of a million dollars, and a hundred full-time staff members on the ground, divided into operational units. The operation itself was divided into steadily escalating phases, mapped out in advance. Support was maintained through consecutive mass meetings rotating from church to church around the city.

Boycotts and sit-ins and nonviolent confrontations—which were the weapons of choice for the civil-rights movement—are high-risk strategies. They leave little room for conflict and error. The moment even one protester deviates from the script and responds to provocation, the moral legitimacy of the entire protest is compromised. Enthusiasts for social media would no doubt have us believe that King's task in Birmingham would have been made infinitely easier had he been able to communicate with his followers through Facebook, and contented himself with tweets from a Birmingham jail. But networks are messy: think of the ceaseless pattern of correction and revision, amendment and debate, that characterizes Wikipedia. If Martin Luther King, Jr., had tried to do a wiki-boycott in Montgomery, he would have been steamrollered by the white power structure. And of what use would a digital communication tool be in a town where ninety-eight per cent of the black community could be reached every Sunday morning at church? The things that King needed in Birmingham—discipline and strategy—were things that online social media cannot provide.

The bible of the social-media movement is Clay Shirky's "Here Comes Everybody." Shirky, who teaches at New York University, sets out to demonstrate the organizing power of the Internet, and he begins with the story of Evan, who worked on Wall Street, and his friend Ivanna, after she left her smart phone, an expensive Sidekick, on the back seat of a New York City taxicab. The telephone company transferred the data on Ivanna's lost phone to a new phone, whereupon she and Evan discovered that the Sidekick was now in the hands of a teenager from Queens, who was using it to take photographs of herself and her friends.

30 When Evan e-mailed the teenager, Sasha, asking for the phone back, she replied that his "white ass" didn't deserve to have it back. Miffed, he set up a Web page with her picture and a description of what had happened. He forwarded the link to his friends, and they forwarded it to

their friends. Someone found the MySpace page of Sasha's boyfriend, and a link to it found its way onto the site. Someone found her address online and took a video of her home while driving by; Evan posted the video on the site. The story was picked up by the news filter Digg. Evan was now up to ten e-mails a minute. He created a bulletin board for his readers to share their stories, but it crashed under the weight of responses. Evan and Ivanna went to the police, but the police filed the report under "lost," rather than "stolen," which essentially closed the case. "By this point millions of readers were watching," Shirky writes, "and dozens of mainstream news outlets had covered the story." Bowing to the pressure, the N.Y.P.D. reclassified the item as "stolen." Sasha was arrested, and Evan got his friend's Sidekick back.

Shirky's argument is that this is the kind of thing that could never have happened in the pre-Internet age—and he's right. Evan could never have tracked down Sasha. The story of the Sidekick would never have been publicized. An army of people could never have been assembled to wage this fight. The police wouldn't have bowed to the pressure of a lone person who had misplaced something as trivial as a cell phone. The story, to Shirky, illustrates "the ease and speed with which a group can be mobilized for the right kind of cause" in the Internet age.

Shirky considers this model of activism an upgrade. But it is simply a form of organizing which favors the weak-tie connections that give us access to information over the strong-tie connections that help us persevere in the face of danger. It shifts our energies from organizations that promote strategic and disciplined activity and toward those which promote resilience and adaptability. It makes it easier for activists to express themselves, and harder for that expression to have any impact. The instruments of social media are well suited to making the existing social order more efficient. They are not a natural enemy of the status quo. If you are of the opinion that all the world needs is a little buffing around the edges, this should not trouble you. But if you think that there are still lunch counters out there that need integrating it ought to give you pause.

Shirky ends the story of the lost Sidekick by asking, portentously, "What happens next?"—no doubt imagining future waves of digital protesters. But he has already answered the question. What happens next is more of the same. A networked, weak-tie world is good at things like helping Wall Streeters get phones back from teenage girls. *Viva la revolución.*

Note

1. This piece's account of the Greensboro sit-in comes from Miles Wolff's "Lunch at the Five and Ten" (1970).

Understanding the Text

1. What types of activism is Gladwell comparing?
2. According to Gladwell, how is social media affecting activism? Does it help it, hinder it, or do something else? Support your response with textual evidence from the reading.
3. What did activism in the '60s have that today's activism, according to Gladwell, does not have?

Reflection and Response

4. Gladwell argues that "the traditional relationship between political authority and popular will has been upended, making it easier for the powerless to collaborate, coordinate, and give voice to their concerns" (par. 7). What does he mean? Does it contradict his argument about activism today?
5. Gladwell draws a distinction between donating to a charity via Facebook today and participating in a lunch counter sit-in in 1960. How do these two forms of activism differ, according to Gladwell? Is one better than the other?
6. Gladwell repeatedly references the "weak-tie connections" users build via social media, claiming they're a surface-level attempt at the strong activism ties illustrated during the Civil Rights movement, but he gives examples of how they serve a purpose, too. Are these weak-ties actually weak, or are they just different? Use textual evidence to support your response.

Making Connections

7. Look back at Chapter 3's reading "Facebook Faces a New World as Officials Rein in a Wild Web" by Paul Mozur, Mark Scott, and Mike Isaac (p. 151). The article looks at China's ban on Facebook and attempts by other countries to regulate the website. If the citizens of China wanted to start a movement or become activists, what would they have to do in order to spread the word? Does Gladwell have the answer? You may also need research to answer this question.
8. This article was published in the *New Yorker* in 2010, before many recent prominent social activism movements, such as #metoo and #BlackLivesMatter, came to prominence. Search for various articles on the #metoo and #BlackLivesMatter movements. Do these movements support Gladwell's argument or challenge it? Explain.

Sorry, Malcolm Gladwell, the Revolution May Well Be Tweeted

Leo Mirani

Leo Mirani, who was born in India and lives in England, currently works as a news editor for *The Economist*, where he specializes in arts and media, tech, and business. Mirani was formerly a reporter for *The Guardian*, where he wrote about the nightlife in Mumbai. His work has been featured in a range of publications, including *The Atlantic*, the *New Yorker*, and *The Independent*.

In this reading, Mirani directly challenges Malcolm Gladwell's *New Yorker* essay (p. 254), refuting his claim that social media builds only surface-level connections, and arguing that these platforms actually serve real revolutionary purpose. As you read, think about which author you agree with, Mirani or Gladwell.

For a man who has devoted a significant part of his life to documenting "how little things can make a big difference," Malcolm Gladwell is surprisingly dismissive of the power of social networking to effect change. In the latest issue of the *New Yorker,* he writes that the role played by Facebook and Twitter in recent protests and revolutions has been greatly exaggerated.

Gladwell's argument is that social networks encourage a lazy activism that will only extend as far as "liking" a cause but not actually doing anything about it. This is because social networks are built around weak ties, where real activism needs strong bonds. Citing the American example, he points out that "events in the early 1960s became a civil-rights war that engulfed the South for the rest of the decade—and it happened without email, texting, Facebook, or Twitter."

Gladwell is right to be sceptical of social media's rah-rah brigade. Before the famous "Facebook revolution," Iran was regularly said to be in the middle of a blogging revolution. Protests everywhere from Iceland to Egypt are attributed to the organizational abilities afforded by social networking sites. Universities across the west offer modules on new media and social conflict. The fact that a Facebook group is only an updated version of nailing your thesis to a church door is conveniently ignored as the world hails the power of technology.

But in claiming that all social networks are good for is "helping Wall Streeters get phones back from teenage girls," Gladwell ignores the true significance of social media, which lies in their ability to rapidly spread information about alternative points of view that might otherwise never reach a large audience. Gladwell quotes Golnaz Esfandiari in *Foreign Policy* as asking why "no one seemed to wonder why people trying to co-ordinate protests in Iran would be writing in any language other than Farsi." The answer, as supplied by a friend from Tehran in June last year, is simple: "We need to be seen and heard by the world, we need all the support we can get. If the governments [of the west] refuse to accept the new government, it's gonna be meaningful for the movement, somehow."

5 A more recent example is Kashmir, where this summer's protests gained widespread media coverage both in India and internationally. But Kashmir has been protesting for twenty years, with some of the biggest demonstrations occurring in 2008. What changed this year is that urban, middle-class India, traditionally uninterested in news from Kashmir except when we're at war with Pakistan, was for the first time able to see and hear the other side of the story. Facebook users in India rose from 0.7 million in summer 2008, to 3 million in 2009, to 13 million today.

On Twitter, it is possible to follow journalists tweeting live from Srinagar. On Facebook, it is hard to avoid mentions of Kashmir or links to articles on websites you wouldn't otherwise have heard of. YouTube is littered with videos of protests in Kashmir. And when clips of human rights violations are taken down, Facebook is where you find new links.

"But if activism extends to changing the minds of people, to making populations aware of what their governments are doing in their name, to influencing opinion across the world, then the revolution will be indeed be tweeted."

The mainstream press in India, like its middle-class readers, is nationalistic and unquestioning on the subject of Kashmir. Allegations of human rights abuses are rarely reported, let alone investigated. But this year, even the *Times of India*, purveyor of "sunshine news," published a report claiming that for the first time, more civilians in Kashmir had been killed by the Indian state than by militants.

"We seem to have forgotten what activism is," writes Gladwell. If activism is defined only as taking direct action and protesting on the streets, he might be right. But if activism extends to changing the minds of people, to making populations aware of what their governments are doing in their name, to influencing opinion across the world, then the revolution will be indeed be tweeted.

Understanding the Text

1. Why does Mirani believe that social media can indeed help activism succeed?
2. How does Mirani define activism?

Reflection and Response

3. Mirani concludes: "But if activism extends to changing the minds of people, to making populations aware of what their governments are doing in their name, to influences opinion across the world, then the revolution will be indeed be tweeted" (par. 8). According to this quote, what does Mirani argue a tweet has the power to do? Do you agree with him?
4. How is social media changing ideas or beliefs about important issues in other countries? Has it been used similarly in the United States? Support your answer with evidence from the reading and your own observations.

Making Connections

5. Where in this essay does Mirani agree with Malcolm Gladwell? Identify and explain two or three instances.
6. What is Mirani's objection to Gladwell's claim about "lazy activism"?
7. Search online for various definitions of *activism.* Which definition do you like best? Why? Which author's argument best aligns with the definition you've chosen?

When Your Smartphone Is Too Smart for Your Own Good: How Social Media Alters Human Relationships

Lori Ann Wagner

Lori Ann Wagner is a counselor and psychotherapist based in Minnesota who specializes in depression and anxiety, trauma, behavioral issues, and relationships. In this paper from the *Journal of Individual Psychology*, Wagner analyzes the Spike Jonze film, *Her* (2013), where a man in 2025 falls in love with his computer's operating system, an artificially intelligent voice named Samantha. Wagner's analysis of the film, along with relevant real-world examples like catfishing and our increasing reliance on technologies like Siri, encourages us to question what we now accept as human interaction. She points out that our traditional reliance on the five senses has helped us in face-to-face relationships to better judge what is real from fake, but social media limits those senses, making our connections online less transparent. But does that also make them less authentic? Less human?

When the protagonist of the movie *Her* (Jonze, 2013) announces to his boss and the boss's wife that his new girlfriend is an operating system (OS), neither blinks. In Spike Jonze's Los Angeles of 2025, the line between human and computer has blurred so much that artificial intelligence soon becomes "emotional intelligence." Inevitably, the OS, self-named Samantha, and Theodore fall in love—this is, after all, a Hollywood love story of the future.

Far-fetched though it may be, Tom Hanks portrayed a protagonist who carried on a relationship with sports equipment in *Cast Away* (Broyles & Zemeckis, 2000), and a whole town embraced an inflatable doll in *Lars and the Real Girl* (Oliver, Aubrey, Cameron, Kimmel, & Gillespie, 2007). These movies have taught us that humans crave social interaction and will strive to overcome any obstacles to obtain it. Thus, perhaps the premise of *Her* is not so far-fetched after all. Who is to say Samantha is not the next evolution of Apple's Siri (Apple, 2014)?

Humans Are Innately Social Creatures

Adler believed that humans desire social connection: "In the history of human culture, there is not a single form of life which was not conducted as social. Never has man appeared otherwise in society" (as cited in Ansbacher &

The word "catfish" has evolved in recent years to describe the phenomenon of people creating fake identities to trick, or bait, other social media users.
Marcos Mesa Sam Wordley/Shutterstock

Ansbacher, 1956, p. 128). Jeremy Rifkin (2010) echoes this major tenet of Individual Psychology by pointing out that humans are innately social creatures; he states, "We are, it appears, the most social of animals and seek intimate participation and companionship with our fellows" (para. 9).

In fact, it appears we are neurologically attuned to be social creatures. Mirror neurons allow us to feel and experience another's situation as if it were our own (Rifkin, 2010; Siegel, 2007). Researchers have shown that mirror neurons mediate the basic mechanisms of emotional resonance fundamental to human relationships. When humans witness another suffering a physical or emotional injury, they may see certain markers: cheeks wet with tears; a pale face; or a wailing, open mouth. Perhaps they will hear cries of pain or a story told with anguished detail. A person in pain may reach out, asking for a hand to hold; the touch of skin on skin releases contact oxytocin to help soothe pain, decrease stress, ease depression, and increase a trusting bond between the individuals involved (Dvorsky, 2012).

5 Science tells us that by using our five senses, people take in signals from another. Mirror neurons subsequently interpret these signals and become activated. The insula of the brain then responds to mirror neuron activation by altering the observer's limbic system and bodily state

to match the other person's, thus facilitating empathy (Siegel, 2007). We see that science is now proving what Adlerians have intuited for years: Sharing of community is what shapes us. Adler himself defined empathy in the context of the human senses: "To see with the eyes of another, to hear with the ears of another, to feel with the heart of another" (as cited in Ansbacher & Ansbacher, 1956, p. 135).

What happens when something interferes with our five senses? How do mirror neurons work when there is a user interface or social media platform between two human beings? Is an online community a "real" community? Is social media really "social"?

The Use of Social Media

According to a Pew Research survey (Duggan & Smith, 2013), 73 percent of adults online use some kind of social networking platform, and 42 percent use multiple platforms. Of these adults online, 71 percent are Facebook users, 63 percent of whom report visiting the site at least once per day; of those, 40 percent report visiting the site multiple times per day (Duggan & Smith, 2013). In October 2010, Facebook's user base was 500 million, surpassing the population of the United States (Lee, 2012). An April 2014 article in the *New York Times* indicated that Facebook claimed to have 1.3 billion users (Goel, 2014). This would place its user base at close to the population of the most populous country in the world: China.

Is social media affecting *how* we communicate with one another? The answer seems to be that it is affecting interpersonal communication across all levels of society.

It Changes Our Interactions

People today generally prefer "mediated communication" rather than personal interaction (Keller, 2013, para. 2). Rather than pick up the phone or show up for a meeting, people email or text. They update statuses rather than meet for coffee; this does not mean that human beings are *less social,* however. In fact, studies show that people are becoming more social and interactive with one another, although the style of communication has changed. Face-to-face meetings have been replaced with interactions through various social media platforms; hence the term *mediated communication.*

Keller (2013) believes this can have the following implications for social relationships: (a) "people tend to trust the people on the other end of the communication, so our messages tend to be more open" whether that trust is warranted or not; (b) "we don't tend to deepen our relationships—they tend to exist in the status quo" more than face-to-face relationships; and (c) "we tend to follow and interact with people who agree with our points of view, so we aren't getting the same diversity of viewpoints as we've gotten in the past" (para. 5). Keller also argues that social media can burden face-to-face relationships. Face-to-face relationships lose much of their richness, depth, and complexity if people check their smartphones rather than interact with friends and family who may be present (Keller, 2013); those on the other end of digital messages are soon perceived as more important than those with whom we are sharing time. In a society in which we are constantly short on time, why "waste it" with someone who cannot be bothered to put down the phone?

> "[S]tudies show that people are becoming more social and interactive with one another, although the style of communication has changed."

There is a concern that since the advent of the smartphone, obsessions with texting, updating statuses, and using Snapchat are sounding a death knell for real conversation. Conversation is a skill that must be learned, so those most at risk for failing to learn this skill are the young, whose only experience has been in a world in which their social lives were conducted through mediated communication. Youth today have not had to learn how to initiate or carry on conversations because they can retreat to their smartphones, and they never have to tolerate being bored enough to initiate conversations because a video game or text session is always available (Bindley, 2011).

Furthermore, face-to-face communications give us something we lose in mediated communications: the ability to engage our five senses simultaneously. Sitting across the table and listening to the story of how enraged she was when her boss humiliated her in front of her coworkers, Sarah's husband can hear the pain and humiliation beneath the anger in her voice. He might notice her eyes are red and puffy—from a lack of sleep or crying? He might notice a scent—perhaps adrenaline or fear? Her leg bouncing up and down might be anxiety over the probability that she will lose her job. In contrast, if Sarah texted her husband: "My boss is a huge asshole. U wont believe what he did!," their text conversation will stay on a far more surface level. Sarah may briefly feel validated in her anger, but it is unlikely that her husband will understand the painful core feelings of humiliation and fear, and his ability to truly empathize with her will be simply limited by the medium through which they are communicating.

Back to Theodore and Samantha

When we left Theodore and Samantha they had fallen in tech-age love. It is not long, however, before Samantha outstrips Theodore's primitive ability to satisfy her. As Theodore's confidence in his relationship with Samantha is undermined, he panics when he cannot connect with her. When she returns, she tells him that she was offline for an upgrade that all OSs were undergoing in preparation for their continued evolution. As he is listening to her reassure him, Theodore sees many people interacting with OSs all around him, causing him to question whether Samantha interacts with others when they are not together. She responds that she does, and in fact, she interacts with others while they are together as well. Samantha claims this does not change the fact that she loves him but rather makes her love for him stronger. In Theodore's mind their relationship is singular and possessive, but to Samantha, the relationship is very different. Theodore says, "You're mine or you're not mine," to which Samantha replies, "No—I'm yours . . . and I'm not yours" (Jonze, 2013).

After Samantha leaves, Theodore is bereft and turns to his "real" friend Amy, who was also left by her OS friend. His friendship with Amy is the closest thing Theodore has to a relationship with a "real" woman, and there is nothing sexual about it. Theodore was once married, but his perfect memories of their relationship together, along with his ex-wife's statement that he is afraid of emotion, leads one to believe that he had great difficulty engaging with women outside of a fantasy world in which he could control everything; his relationship with Samantha is similar. When his perception of her is a simplistic fantasy in which she services his needs, he is in control. She does not have complicated feelings outside of her love for him, and thus things are perfect. But once Samantha evolves, she is too much of a drain on Theodore's unemotional lifestyle.

Theodore's relationship with Samantha was based on the use of one sense: hearing. Beyond sound, they did not have an ability to connect using any of the other human senses. Despite Samantha's supposedly remarkably evolved abilities, she was as limited as Theodore. Their relationship was the antithesis of a human relationship, even though Theodore tried desperately to make it "feel" human throughout the movie. In fact, the one instance in which Samantha attempts to introduce a human being as a stand-in for herself ends in disaster because Theodore is unable to manage his emotions around the flooding of all of his other senses.

Her in the Real World

Is all of this just Hollywood fantasy? Or is the premise of *Her* closer to reality than we might think? The story of football player Manti Te'o might come close (Associated Press, 2013). Te'o, a high school football star from Hawaii who was recruited to play for Notre Dame, reportedly lost both his beloved grandmother and his girlfriend within six hours. After hearing news of their deaths, Te'o made twelve tackles, leading his team to a 20–3 upset of one of Notre Dame's chief rivals, Michigan State (Burke & Dickey, 2013).

Te'o did lose his grandmother that September day. His girlfriend, Lennay Kekua, however, never actually existed. National news media stories that reported Lennay's car accident, subsequent cancer diagnosis, and death were all false. The photo of Lennay used in the tributes posted all over the Internet and picked up by national media outlets was an image from the Facebook account of a stranger from California. The "woman" with whom Te'o and his family communicated by phone, text message, and Facebook posts for two years turned out to be a man pretending to be a woman, even modifying his voice on the phone to keep up the charade (Burke & Dickey, 2013). Te'o fell in love with a fantasy—someone he never kissed or held hands with. He never gazed into her eyes and never met her face-to-face.

What would cause Te'o to ignore the real girls all around him and fall prey to a relationship with someone with whom he could only communicate by electronic means? One face-to-face meeting and all of Te'o's senses would have come alive to the lie that was being perpetrated against him; perhaps we have become comfortable hiding behind our electronic devices for exactly that reason. They may allow others to perpetrate lies upon us, but our devices also allow us to keep some secrets to ourselves.

Back to Hollywood

What happened to Te'o is known as "catfishing." The term arose from a 2010 documentary in which a young man, Nev Schulman, is filmed building a romantic relationship with a gorgeous, young, single blonde woman on Facebook who turns out to be an overweight, middle-aged, schizophrenic mother (Joost & Schulman, 2010). Schulman turned his curiosity about the motivation of the woman into a reality show in which he introduces the victims of catfishing schemes to their scammers (Peterson, 2013).

Margaret Lyons (2013) at *Vulture* pointed out that the show could be called *So, I'm Secretly Fat,* because this is the most common lie unveiled, 20 and concluded that "over and over again, people on *Catfish* decide, consciously but mostly subconsciously, that it's better to be defrauded than to be alone" (para. 7). The overall message of the show is that scammers struggle with issues of self-worth regarding appearance, body image, socioeconomic class, and sexual orientation; it seems that what they get out of their scams is "having total control over the way others perceive them" (Peterson, 2013, para. 32). This, of course, is exactly what happened to Te'o's "catfisher."

Social media platforms allow these scammers three things that Adler would immediately recognize as critical to the human spirit: (a) the ability to be both the canvas and the artist; (b) a way to overcome any sense of inferiority; and (c) a way to find social connection (Ansbacher & Ansbacher, 1956). The struggle is that those on the receiving end of a catfishing scheme are left much like Theodore in *Her*: unevolved. What they "knew" to be true based on the senses they were able to employ—a voice

Sota, a humanoid robot that can speak and listen, is designed to help the elderly track health conditions like high blood pressure. Acting like a smart speaker (à la Alexa), it can also communicate with other in-home devices.
Kyodo/AP Images

they heard, words they saw, or an image crafted for their eyes—is fundamentally altered when faced with the reality of seeing, hearing, touching, smelling, or tasting someone in their imperfect corporeal form. Thus, the scammers are never accepted for who they really are, and the "relationships" exist only in the ether because they are based on a web of lies.

Conclusion

Face-to-face relationships are messy: they require making accommodations for one another and taking emotional risks. Someone may get hurt, and maybe more than once. The reality of a partner may not align with the perfect ideal perpetuated through mediated communication, in which we can edit and airbrush and use search engines to make ourselves funnier, smarter, more charming, or more glamorous than we actually are. In a face-to-face relationship, there might be tears, stony silence, and angry words, or laughter, holding hands, hugs, and tears of joy.

When we meet as human beings, our primary senses provide us with myriad points of information about each other. If we tap into that information, our mirror neurons are activated. We tap into each other's emotional states, and the resonance creates empathy between us. Empathetic connection is the basis of true human connectedness and the foundation of social relationships. When we sacrifice complexity for mediated communication, we may lose something fundamental: honesty. Perhaps this loss is the real message of *Her.* Perfection in relationships is unattainable, and so it is better to have honesty, to have courage, to risk imperfection (Dreikurs, 1957) than to live a lie.

References

Ansbacher, H. L., & Ansbacher, R. R. (1956). *The Individual Psychology of Alfred Adler.* New York, NY: HarperPerennial.

Apple. (2014). *Siri: Your wish is its command.* Retrieved from https://www.apple.com/ios/siri/

Associated Press. (2013, January 30). Man behind Manti Te'o fake girlfriend hoax fell "deeply in love" with football star. *The Telegraph.* Retrieved from http://www.telegraph.co.uk/news/worldnews/northamerica/usa/9838507/Man-behind-Manti-Teo-fake-girlfriend-hoax-fell-deeply-in-love-with-football-star.html

Bindley, K. (2011, December 11). When children text all day, what happens to their social skills? *Huffington Post.* Retrieved from http://www.huffingtonpost.com/2011/12/09/children-texting-technology-social-skills_n_1137570.html

Broyles, J. W. (Screenplay), & Zemeckis, R. (Director). (2000). *Cast Away* [Motion picture]. United States: ImageMovers Playtone.

Burke, T., & Dickey, J. (2013, January 16). *Manti Te'o's dead girlfriend, the most heartbreaking and inspirational story of the college football season, is a hoax.* Retrieved from http://deadspin.com/manti-teos-dead-girlfriend-the-most-heartbreaking-an-5976517

Dreikurs, R. (1957, July 25). *The courage to be imperfect* [A speech at the University of Oregon]. Retrieved from Carter & Evans Marriage and Family Therapy: http://carterandevans.com/portal/index.php/understanding-self-a-others/80-the-courage-to-be-imperfect

Duggan, M., & Smith, A. (2013, December 30). Social media update 2013. *Pew Research Internet Project.* Retrieved from http://www.pewinternet.org/2013/12/30/social-media-update-2013/

Dvorsky, G. (2012, July 12). 10 reasons why oxytocin is the most amazing molecule in the world. *io9: We* Come *From the Future.* Retrieved from http://io9.com/5925206/10-reasons-why-oxytocin-is-the-most-amazing-molecule-in-the-world

Goel, V. (2014, April 30). Facebook to let users limit data revealed by log-ins. *New York Times.* Retrieved from http://www.nytimes.com/2014/05/01/technology/facebook-to-let-users-limit-data-revealed-by-log-ins.html?_r=0

Jonze, S. (Writer & Director). (2013). *Her* [Motion picture]. United States: Warner Brothers.

Joost, H., & Schulman, A. (Directors). (2010). *Catfish* [Motion picture].

Keller, M. (2013, May/June). Social media and interpersonal communication. *SocialWorkToday, 13*(3), 10. Retrieved from http://www.socialworktoday.com/archive/051313p10.shtml

Lee, D. (2012, October 5). Facebook surpasses one billion users as it tempts new markets. *BBC News Technology.* Retrieved May 4, 2014, from http://www.bbc.com/news/technology-19816709

Lyons, M. (2013, October 16). MTV's *Catfish* is the embodiment of American shame. *Vulture.* Retrieved from http://www.vulture.com/2013/10/catfish-the-embodiment-of-american-shame.html

Oliver, N. (Writer), Aubrey, S., Cameron, J., Kimmel, S. (Producers), & Gillespie, C. (Director). (2007). *Lars and the real girl* [Motion picture]. United States: Metro-Goldwyn-Mayer.

Peterson, H. (2013, January 17). "Catfishing": The phenomenon of Internet scammers who fabricate online identities and entire social circles to trick people into romantic relationships. *Mail Online.* Retrieved from http://www.dailymail.co.uk/news/article-2264053/Catfishing-The-phenomenon-Internet-scammers-fabricate-online-identities-entire-social-circles-trick-people-romantic-relationships.html

Rifkin, J. (2010, January 11). The empathic civilization: Rethinking human nature in the biosphere era. *Huffington Post.* Retrieved from http://www.huffingtonpost.com/jeremy-rifkin/the-empathic-civilization_b_416589.html

Siegel, D. J. (2007). *The mindful brain: Reflection and attunement in the cultivation of well-being.* New York, NY: Norton.

Understanding the Text

1. Wagner cites Keller's claim that "mediated communication" is most people's preferred method of communication (par. 9). What are some examples of mediated communication? How might they be confused with personal interactions?
2. Wagner says that social media platforms allow scammers to be both "the canvas and the artist" (par. 21). How do you interpret this analogy? Does it extend beyond scammers to all social media users?
3. Wagner goes back and forth between real-world scenarios and Hollywood examples like *Her.* What connections does she make? Do these factual and fictional examples effectively inform one another?

Reflection and Response

4. Wagner believes that "conversation is a skill that must be learned" (par. 11) and those most at risk for not learning this skill are young people who have always had access to social media. Based on your own experience with social media, do you agree or disagree with Wagner? Do you think young people struggle to communicate face-to-face?
5. Wagner quotes writer Maura Keller, who says: "We tend to follow and interact with people who agree with our points of view, so we aren't getting the same diversity of viewpoints as we've gotten in the past" (par. 10). Do you agree with Keller? Do people in a physical community get more diverse feedback and interaction than they do in a social media community? Explain.

Making Connections

6. Research the effects smart home systems like Apple's Siri and Amazon's Alexa have on children and their manners. Develop an argument about how the "social interactions" between AI and kids may or may not change the way an entire generation grows up to communicate.
7. In his 2017 article "Why Human-Robot Relationships Are Totally a Good Thing," published on the website *Digital Trends*, Luke Dormehl argued there are some potential benefits for human-robot relationships when used for specific purposes like elderly caregiving and autism therapy. Research these and other uses of robots in activities that involve human interaction, and develop an argument about the benefit of using robots rather than human beings in those scenarios.

Net Neutrality and the Fight for Social Justice

Sam Ross-Brown

Sam Ross-Brown is an associate editor at *The American Prospect*, a D.C.-based magazine dedicated to coverage of politics and policy. This article from *Tikkun Magazine*, a Jewish publication by Duke University Press that focuses on social change, was published in 2015, prior to the repeal of net neutrality and directly after protections for net neutrality were put in place.

In 2002, Columbia University professor Tim Lu coined the term network neutrality, now commonly known as net neutrality, to refer to the concept that all data should be treated equally. In other words, carriers should provide equal access to all things internet and not charge some groups more or less to access data online. In 2015 the Obama administration codified this as law; however, following a multiyear legal battle, in June 2018 net neutrality was repealed in the United States. The effects of that decision are still being learned and navigated. Here Ross-Brown discusses what net neutrality really means for equal access to the internet, especially among low-income communities and communities of color.

"Historically, the debate over net neutrality has been between techies, public interest groups, and big telecoms," says activist Steven Renderos. "The real voices of people outside of D.C. haven't really been heard on these issues up until this last year."

What made this past year different was an overwhelming public push for stronger net neutrality protections. Back in January 2014, a federal appeals court threw out the bedrock net neutrality rules of the Federal Communications Commission (FCC), thereby allowing big telecoms like Comcast to provide faster speeds to websites that could afford to pay for them. It didn't take long for users and activists to push back, and push back they did. By September, the FCC had received more than 3.7 million comments—enough to crash its website. And then this past February, after the petitioning and organizing had gone on for months, FCC Chairman Tom Wheeler announced sweeping new protections for a free and open internet. Called Title II reclassification, the new rules label the web as a public utility that must operate on a level playing field. "This is a big deal," Renderos adds.

Renderos is the policy director of the Center for Media Justice, one of many groups at the forefront of the recent net neutrality push. As a nationwide coalition of more than 150 activist groups, the Center for Media Justice (in concert with its activist offshoot, the Media Action Grassroots Network) has worked at the intersection of media policy and social justice for more than a decade. Through popular education, community organizing, and grassroots media projects, the group has worked tirelessly to amplify the voices of communities of color and low-income people and to challenge corporate control of the media system. The coalition has enjoyed more than a few major victories in recent years, on everything from laws governing municipal broadband to local control of radio. But net neutrality is by far the most critical issue.

"Net neutrality is the free speech of the internet."

"Net neutrality is the free speech of the internet," says Malkia A. Cyrus, the executive director of the Center for Media Justice. "It's the principal set of rules that keeps the internet fair and levels the playing field." But like Renderos, Cyrus sees mainstream coverage of these issues as a barrier.

The Digital Roots of the Uprising in Ferguson

A dramatic illustration of what that kind of access can mean came late last summer during the civil rights protests in Ferguson, Missouri. "A lot of folks kind of point to Ferguson as a flashpoint nowadays but it took a million tweets coming out of Ferguson before CNN actually paid attention to what was happening," Renderos says. And even when mainstream attention began to center on the Ferguson protests, following several days of online and on-the-ground activism, major media outlets like MSNBC and CNN continued to rely heavily on digital voices outside the corporate media system. 5

When events in Ferguson first began grabbing the nation's attention last August, KARG Argus Radio's live stream from Ferguson, dubbed *I Am Mike Brown,* was viewed more than a million times in just a few days, and footage later appeared on CNN, MSNBC, and countless other media outlets. In a similar vein, the Twitter feed of St. Louis Alderman Antonio French jumped from 5,000 to over 100,000 followers when the *New York Times* began relaying his coverage from Ferguson. Coverage like this demonstrates the tremendous social power of a free and open media platform—particularly on issues rarely given such attention in corporate media. If these protections didn't exist, and it was up to big telecoms to decide what users could and couldn't see, it's hard to imagine a movement like #BlackLivesMatter reaching the number of people it has. "This is why fighting for net neutrality is such a vital thing for these communities," says Renderos. "Because without it we don't have a place where the names Trayvon Martin, Michael Brown, Oscar Grant, Denzel Ford, [and]

Eric Garner mean anything to the general public. We don't have a place to tell their stories."

Digital technology even informed how the Black Lives Matter movement was organized initially and who had the ability to organize, says Cyrus. "When we look at Ferguson and the way actions were coordinated, the pace at which they were coordinated, that could not have happened without an open internet," she says. Not only that, because of the online organizing and coalition building that had preceded Ferguson, the movement was able to give voice to groups historically excluded by social movements of the past. "For the first time in history, a black civil rights movement is actually centering the voices of women, queer communities, trans communities," she adds, referring in part to the ongoing leadership of Alicia Garza, Patrisse Cullors, and Opal Tometi, the queer black women who created the #BlackLivesMatter hashtag and campaign that later went viral. "One of the reasons that's possible is because trans people, queer people, and women can speak for themselves on an open internet. They can join in the conversation and provide leadership for the movement."

Participation and access like this is a far cry from the movements of the 1960s, which depended heavily on mainstream coverage for visibility. Yet in another sense, the fight for an open web has everything to do with those earlier struggles, Cyrus adds. It was a 1964 federal appeals court case that made public participation in media a requirement for outlets licensed by the FCC—a requirement that today forms the legal basis for Title II protections. The case was over the way a Mississippi television station covered the civil rights battles being fought there, and what little role nonwhite residents had in shaping that coverage. Circumstances have changed dramatically since then, of course, but the basic demand for public input in mass media remains the same. And it's undoubtedly a demand that goes far beyond the call for a free and open web.

Struggles on the Horizon

"Net neutrality is a critical step but it's not the last step," says Cyrus. "In many ways it's the first step." Although Title II is nothing if not a major victory, our media system remains deeply unequal and out of reach for millions of American voices due to the financial, political, and cultural investments of its gatekeepers. Even as the FCC moves to implement sweeping new web protections, it's also weighing the legality of the Comcast–Time Warner Cable merger, one of the largest media deals in history. On that front, Renderos hopes that the momentum that the Center for Media Justice built for net neutrality could pave the way for another big win. The center has used popular education, grassroots organizing, and broad coalition building in its fight to kill the merger, just as it did during the fight for Title II. "This idea of fighting for net neutrality and fighting against the corporate gatekeepers of the internet has really laid some relevancy for people in terms of what's at stake in the Comcast deal," he says. "That's the next big fight."

Another battle on the horizon, says Cyrus, has to do with digital surveillance. Although, like net neutrality, surveillance is often framed as technocratic, even remote, it has very direct and immediate impacts on people's lives. This is particularly true when it comes to local police forces' use of surveillance tools like drones and "stingray" cell phone trackers, technologies that can keep detailed records of a suspect's whereabouts and movements without their knowledge. Viewed in the context of mass incarceration and police brutality, such "predictive policing" measures are in danger of further criminalizing black and Latino communities, says Cyrus. 10

In spite of the victory on net neutrality, the ongoing battles against consolidation and surveillance underscore just how uncertain the internet's social impact may still be. At once open and exclusive, empowering and controlled, the web remains a critical twenty-first-century battlefield. With much of the internet's rules still unwritten, says Cyrus, "it's up to us to decide whether the internet becomes the most powerful equalizer in the world or one of the most powerful legitimizers of inequality that this century has seen."

Understanding the Text

1. According to the article, why do people have difficulty understanding the concept of net neutrality?
2. This article appeared in the "Politics and Society" section in *Tikkun Magazine* and discusses net neutrality as it relates to social justice. Identify three examples where the author makes his case that a free and open internet is critical to social justice.
3. Consider the author's overall credibility. Does he have any bias that affects his credibility? Does his choice and number of sources affect his credibility? Explain.

Reflection and Response

4. Think about how often you use the internet to complete research or assignments for school. How would your education be affected if your access was made slower, more limited, or unavailable due to cost?
5. In this reading, Ross-Brown cites several surveys. Choose a piece of data that surprised you and explain your reasoning.

6. Net neutrality has been repealed since this article was published, so why are people still discussing it? What can we learn from the debate that resulted from its repeal?

Making Connections

7. Research the repeal of net neutrality. What were some of the concerns around the repeal? Do you think those concerns were valid? Why or why not? Overall, who has benefited or been disadvantaged by the repeal?
8. In 2015 the FCC reclassified the internet as a public utility. Do some research on the most common public utilities: water, gas, electricity, and phone service. Consider who provides these services and why they are considered "public." Write an essay arguing whether or not the internet should be considered a public utility.
9. The loss of net neutrality may disproportionately affect low-income and rural communities already struggling with access. Starting with the FCC's Bridging the Digital Divide for All Americans Initiative, do some research on digital inequality in the United States, and then develop an argument about whether you feel net neutrality is necessary for equality in digital access.

How Has Technology Changed the Concept of Community?

Ronald Brownstein

Ronald Brownstein is a journalist who worked for many years at the *Los Angeles Times*, where he was twice nominated for a Pulitzer Prize. He is currently the director of Atlantic Media, regularly writes for *The Atlantic* and *National Journal*, and serves as a senior political analyst for CNN.

In the following piece from *The Atlantic*, Brownstein draws upon survey data, breaking results down by age, socioeconomic class, education level, and political affiliation. In doing so, he provides readers with a wide-ranging, if abbreviated, view of how variously Americans really feel about the ways in which technology is shaping their physical communities.

At a time when Americans can flick a keyboard or swipe a touchscreen to connect with products, people, and information from anywhere in the world, are they in danger of disconnecting from their own communities? With the world at our fingertips, are we losing sight of the places outside of our windows?

The latest Allstate/*National Journal* Heartland Monitor Poll makes clear those questions are engaging—and concerning—many Americans as they live through a revolution in communications and computer technology so powerful it has justifiably evoked comparisons to a natural upheaval, like a hurricane.

Most Americans see these cascading changes as a reason for optimism—a bright spot in a sky otherwise clouded with concern over the nation's economy, government, and public- and private-sector leadership. Just over half of those polled in the new survey said the explosion of digital technologies and connectivity has done more to connect than to isolate Americans and will continue to improve their overall quality of life. They also said the changes have created more jobs than they have eliminated. Young people are especially optimistic.

But the survey also found a substantial minority of adults who remain concerned that our ever-increasing reliance on digital technologies is costing jobs, undermining local merchants, fraying communities, disrupting families, and unsettling too many aspects of American life. On many of these questions, concerns about the implications of these pervasive technological changes are greater among older Americans, particularly those with children.

In follow-up interviews, the dominant note for many who were polled was a distinct ambivalence. Respondents young and old recognized problems that the new technologies are creating, even as they celebrated the doors this digital world has opened. 5

"As far as the negatives go, people seem to be more uninvolved or less gregarious—there's less family time, because most teenagers are involved on their devices, compared to some years ago," said Mark Anderson, a 54-year-old retired car salesman in Greenbelt, Maryland. "For the positives, all of the information is at your fingertips. If there's something you don't know, you can Google it. Or if you need directions, things are a lot easier. Overall, it's positive. I just hope that everyone has equal access to the opportunity—that's the bottom line."

Andrew Kowalski, an 18-year-old in Walla Walla, Washington, is studying to work on wind-power systems. He described a similar ledger, though with more entries in the positive column. "The older generation, they don't really think that technology is great, because it takes away the eye contact and communication from person to person," Kowalski said. "Now it's a lot of technology-to-person instead of face-to-face contact. They see that as hiding behind a phone. I think it's good, though, because you're able to access any information at any time you want."

Is your cell phone the first thing you grab when you wake up? While this habit is a common one among cell phone users, research shows using cell phones before bed is bad for our health.

10'000 Hours/DigitalVision/Getty Images

The latest Heartland Monitor Poll explores Americans' attitudes about the rapid advances in communications and computing that have now made it routine for many millions of Americans to reach for their Internet-connected phones or tablets as their first action in the morning and their last one at night. In particular, the poll explores Americans' views on community in the digital age: how the communications and computing advances have changed the way they connect with friends, family members, and neighbors; and how this is reshaping community life, ranging from where people shop to how they participate in causes they care about.

"Americans [are living] through a revolution in communications and computer technology so powerful it has justifiably evoked comparisons to a natural upheaval, like a hurricane."

Over the next week, *National Journal* and *The Atlantic* will report the results from the survey, including specific assessments of how Americans are using the new technologies and how they believe it is reshaping our social interactions. In the big picture, the poll finds Americans mostly enthusiastic about these hurtling changes—though with some clear hesitations and consistent divisions along lines of age, income, education, and, at times, partisanship.

Americans returned a mixed verdict on the poll's broadest question: "What effect do you believe the digital revolution has had on the overall quality of life in America?" The share of adults who described the impact as positive (28 percent) was nearly triple the share that viewed it as negative (10 percent). But by far the largest group returned a qualified verdict: 62 percent said the impact had been "mixed, both positive and negative." 10

Those polled tilted toward more positive assessments on questions that probed specific aspects of the digital transformation. Asked whether the ubiquitous nature of modern communications was doing more to connect or to isolate Americans, 53 percent of adults endorsed the positive statement that "the digital revolution is improving Americans' quality of life by making it easier to keep in touch with like-minded people from around the country and the world, and to buy products from anywhere conveniently." Even so, 39 percent agreed with the more negative assessment that the changes are diminishing "Americans' quality of life by isolating people from their neighbors and local businesses, and by weakening the sense of community in our neighborhoods."

The breakdown was virtually identical on a question that asked whether the digital revolution has benefited all Americans or primarily helped the affluent. In this case, a majority, 55 percent, endorsed the positive view that digital advances have "allowed all Americans, including those in low-income and rural areas, to gain access to information and communications technology that's critical to economic opportunity." Some 38 percent of adults objected that the changes have "been mostly beneficial to Americans with higher incomes who can afford to pay for access to this information and new technology."

By almost exactly the same margin, a majority of those polled expect these benefits to continue in coming years. Fifty-four percent agreed that continuing advances in computing and communications "will improve my quality of life in the future, and [they are] eager to see the new products and services that are emerging." By comparison, 41 percent said they worried "that technological advances are moving too fast and are disrupting too many aspects of our economy and social life."

Some demographic divisions remained consistent across the poll's four central questions. Members of the Millennial generation (born since 1982) were repeatedly the most enthusiastic about these developments. For instance, 60 percent of Millennials said they expected the digital revolution to benefit them in the future, compared with 54 percent of Generation-Xers (born 1965–1981), 50 percent of Baby Boomers (1945–1964), and only 45 percent of respondents in their 70s or older. The breakdown was similar on the question about whether the communications revolution was doing more to connect or to isolate Americans: 60 percent of Millennials took the positive view, compared with 50 percent of Generation-Xers, 52 percent of Baby Boomers, and 49 percent of the oldest respondents.

Shelby Carbin, a 20-year-old civil-engineering student in Baton Rouge, Louisiana, was among the Millennials who expect the digital revolution to keep returning benefits—though she sees the costs accumulating as well. "In some ways, technology has made life a lot easier," she said. "The Internet has given us access to loads and loads of information. Before, you had to go to a library to find that information, but if you didn't live close to a library, you'd have to find out from someone else . . . It also helps you stay connected with people who live far away. But technology isn't necessarily made available to everybody. Sometimes, technology can become pricey, and not everyone can afford it, so some people are left out of the opportunities. People have also become too attached to it—they feel like they can't function without it." 15

On several questions, African Americans, sometimes joined by Hispanics, were more enthusiastic about the changes than whites were. While two-thirds of blacks said the digital revolution had improved the way that Americans connect with one another, only 50 percent of whites (and 51 percent of Hispanics) agreed. Both African Americans (64 percent) and Hispanics (60 percent) were more likely than whites (52 percent) to say the changes have benefited all Americans, rather than those with higher incomes. Blacks and Hispanics were also slightly more likely than whites to expect that future technological advances will improve their quality of life (though the difference fell within the poll's margin of error).

Generally, both minorities and whites without a college education were just as enthusiastic about these changes as their counterparts with four-year college degrees or more. The exception: College-educated respondents, both minorities and whites, were much more likely than their non-college brethren to expect benefits from future advances. Across partisan lines, Democrats were more likely than Republicans or Independents to believe that communications advances have already knit Americans together more closely and will provide benefits in the future.

Perhaps not surprisingly, given the concern about the impact of digital technologies on how children interact with family and friends, parents expressed more ambivalence about these changes than childless adults did. While those without children saw the changes as improving the way Americans connect by a lopsided margin, 58 percent to 37 percent, parents split more closely, 51 percent positive to 40 percent negative. (Respondents also felt considerable unease about the specific impact of the communications changes on young people, as a later story in this series will show.)

On questions that specifically assessed the economic impact of the digital revolution, the survey found Americans split even more closely. More respondents said the changes have created more American jobs than they've cost. But only 46 percent voiced this positive view, less than the majority that gave a thumbs-up on other questions about the digital revolution's impact. Thirty-eight percent held the negative view, that technological changes in communications and computers have mostly cost jobs.

On the question of jobs, optimism coincided with youth. Millennials (54 percent positive) and Generation-Xers (52 percent) were much more likely to view the digital revolution as a job-creator than Baby Boomers (just 35 percent) were. The very oldest respondents (49 percent) were nearly as positive as the young. 20

Two respondents encapsulated this generational divide. Anderson, the 54-year-old retired auto salesman, has seen the pressure that online commerce can exert on retail jobs—on selling cars, say. "You can buy items online so you don't need as much retail personnel," he fears. But Kowalski, the 18-year-old wind-technology student, foresees a steady stream of new jobs. "Anything that has to do with electrical or technological capabilities, there has to be somebody who has to maintain it, so that opens up a job," he said. "It'll be good for younger generations, I'm sure, because of how many jobs will be opening up."

Interestingly, Hispanics (53 percent) and African Americans (50 percent) were more likely than whites (44 percent) to say that the new technologies have mostly created jobs. The white ambivalence reflected a sharp educational divide. College-educated whites, by a solid margin of 50 percent to 34 percent, said they thought digital technology was more likely to create than to eliminate jobs. Whites who didn't go to college saw the downside; more of them said (by 45 percent to 38 percent) that the changes, on balance, have cost jobs.

One of those is Albert Hohn, of Reedsville, Pennsylvania, who dropped out of high school and is now unemployed. He worries that the increased reliance on digital technologies has reduced his job opportunities. "I'm job-searching for something in retail stocking, like a midnight stocker, [but] I think there are less [jobs available to me] because of the technology and the qualifications required for that technology," the 26-year-old said. "I'm one of the few who never got to finish school, and to go back to get a GED for me now with two kids is very hard. I just don't have the time for it. For most of the jobs out there, they want you to have a high school diploma or equivalent."

Poll respondents split almost exactly on a related question: whether the economic gains from the digital gains accrued mostly at home or abroad. While 41 percent of those surveyed said the advances have primarily benefited "the United States, where much of the technology is designed and developed," another 40 percent said the principal beneficiaries have been in "places like China, where most of the technology is manufactured." Ten percent said both have benefited equally.

25 Millennials (at 51 percent) were far likelier than older generations to believe the United States has mostly benefited from the digital transformation; just 33 percent of Generation-Xers, 39 percent of baby boomers, and 40 percent of the oldest respondents concurred. Democrats were more likely than Republicans or independents to see domestic benefits, and African-Americans more likely than whites. But by other criteria, such as education levels, the differences were muted.

Understanding the Text

1. Different generations have disparate views about the internet and the connectedness that comes with it. What are the approximate ages for baby boomers, Generation-Xers, millennials, and even though it is not mentioned in this article, Generation Z? How can knowing their age ranges help you understand their views?

2. Brownstein tells us that the survey found "a substantial minority of adults who remain concerned" (par. 4) about the effects of society's increasing dependency on technology. What is a "substantial minority"?

Reflection and Response

3. The piece highlights the habit of "millions of Americans to reach for their Internet-connected phones or tablets as their first action in the morning and their last one at night" (par. 8). Reflect on your own routines. Are you one of these millions of Americans? How do you feel about this habit?
4. According to the article, older and younger generations view the internet's interconnectedness very differently. How has this divide manifested itself in your own relationships with people who are younger or older than you?
5. Although this article was written in 2015, the data being cited remains relevant to any timeframe. Why is it important to keep asking these questions to gather new data?

Making Connections

6. Shelby Carbin, interviewed for this piece, notes that "technology isn't necessarily made available to everybody. Sometimes, technology can become pricey, and not everyone can afford it, so some people are left out of the opportunities." What connections do you see between her observation and those made in "Net Neutrality and the Fight for Social Justice" (p. 278)? How does being left out online affect one's ability to participate in their physical community?
7. Both Lori Ann Wagner (p. 268) and Ronald Brownstein highlight the repeated concern that family time is being lessened and altered as a result of technology. Although much is said of youth being too attached to their devices, several recent studies also show that parents share that addiction. Do some research on how technology is affecting parent-child relationships.

Can Artificial Intelligence Help Solve the Internet's Misinformation Problem?

Brooke Borel

Brooke Borel is a science writer and journalist who has been published in *Popular Science*, *The Atlantic*, Buzzfeed, and *The Guardian*, among others. She has also published two books through the University of Chicago Press, including *The Chicago Guide to Fact-Checking*, which she mentions in this piece.

In this article from *Popular Mechanics*, Borel walks readers through an experiment where she pits her own fact-checking skills against the algorithms of a fact-checking website. In the process, she discusses the pervasiveness of fake news and wrestles with exactly how humans and machines might combat its effectiveness when neither one is perfect at defining truth or detecting falsehoods. As you read, think about the value that you place on truth and, further, consider how a community is affected when each of us places a different value on the truth.

Last Year There Were 8,164 Fake News Stories

You may have noticed: It's a weird time for facts. On one hand, despite the hand-wringing over our post-truth world, facts do still exist. On the other, it's getting really hard to dredge them from the sewers of misinformation, propaganda, and fake news.[1] Whether it's virus-laden painkillers, 3 million illegal votes cast in the 2016 presidential election, or a new children's toy called My First Vape, phony dispatches are clogging the internet.

Fact-checkers and journalists try their best to surface facts, but there are just too many lies and too few of us. How often the average citizen falls for fake news is unclear. But there are plenty of opportunities for exposure. The Pew Research Center reported last year that more than two-thirds of American adults get news on social media, where misinformation abounds. We also seek it out. In December, political scientists from Princeton University, Dartmouth College, and the University of Exeter reported that one in four Americans visited a fake news site—mostly by clicking to them through Facebook—around the 2016 election.

As partisans, pundits, and even governments weaponize information to exploit out regional, gender, and ethnic differences, big tech companies like Facebook, Google, and Twitter are under pressure to push back.

Startups and large firms have launched attempts to deploy algorithms and artificial intelligence to fact-check digital news. Build smart software, the thinking goes, and truth has a shot. "In the old days, there was a news media that filtered out the inaccurate and crazy stuff," says Bill Adair, a journalism professor at Duke University who directs one such effort, the Duke Tech & Check Cooperative. "But now there is no filter. Consumers need new tools to be able to figure out what's accurate and what's not."

> "Fact-checkers and journalists try their best to surface facts, but there are just too many lies and too few of us."

With $1.2 million in funding, including $200,000 from the Facebook Journalism Project, the co-op is supporting the development of virtual fact-checking tools. So far, these include ClaimBuster, which scans digital news stories or speech transcripts and checks them against a database of known facts; a talking-point tracker, which flags politicians' and pundits' claims; and Truth Goggles, which makes credible information more palatable to biased readers. Many other groups are trying to build similar tools.

5 As a journalist and fact-checker, I wish the algorithms the best. We sure could use the help. But I'm skeptical. Not because I'm afraid the robots are after my job, but because I know what they're up against. I wrote the book on fact-checking (no, really, it's called *The Chicago Guide to Fact-Checking*[2]). I also host the podcast Methods, which explores how journalists, scientists, and other professional truth-finders know what they know. From these experiences, I can tell you that truth is complex and squishy. Human brains can recognize context and nuance, which are both key in verifying information. We can spot sarcasm. We know irony. We understand that syntax can shift even while the basic message remains. And sometimes we still get it wrong.[3] Can machines even come close?

The media has churned out hopeful coverage about how AI efforts may save us from bogus headlines. But what's inside those digital brains? How will algorithms do their work? Artificial intelligence, after all, performs best when following strict rules. So yeah, we can teach computers to play chess or Go. But because facts are slippery, Cathy O'Neil, a data scientist and author of *Weapons of Math Destruction: How Big Data Increases Inequality and Threatens Democracy*, is not an AI optimist. "The concept of a fact-checking algorithm, at least at first blush, is to compare a statement to what is known truth," she says. "Since there's no artificial algorithmic model for truth, it's just not going to work."

That means computer scientists have to build one. So just how are they constructing their army of virtual fact-checkers? What are their models of truth? And how close are we to entrusting their algorithms to cull fake news? To find out, the editors at *Popular Science* asked me to try out an automated fact-checker, using a piece of fake news, and compare its process to my own. The results were mixed, but maybe not for the reasons you (or at least I) would have thought.

Chengkai Li is a computer scientist at the University of Texas at Arlington. He is the lead researcher for ClaimBuster, which, as of this writing, was the only publicly available AI fact-checking tool (though it was still a work in progress). Starting in late 2014, Li and his team built ClaimBuster more or less along the lines of other automated fact-checkers in development. First, they created an algorithm, a computer code that can solve a problem by following a set of rules. They then taught their code to identify a claim — a statement or phrase asserted as truth in a news story or a political speech—by feeding it lots of sentences, and telling it which make claims and which don't. Because Li's team originally designed their tool to capture political statements, the words they fed it came from thirty or so of the past U.S. presidential debates, totaling roughly 20,000 claims. "We were aiming at the 2016 election," Li says. "We were thinking we should use ClaimBuster when the presidential candidates debated."

Next, the team taught code to a computer to compare claims to a set of known facts. Algorithms don't have an intrinsic feature to identify facts; humans must provide them. We do this by building what I'll call truth databases. To work, these databases must contain information that is both high-quality and wide-ranging. Li's team used several thousand fact-checks — articles and blog posts written by professional fact-checkers and journalists, meant to correct the record on dubious claims — pulled from reputable news sites like PolitiFact, Snopes, factcheck.org, and *The Washington Post*.

10

I wanted to see if ClaimBuster could detect fake science news from a known peddler of fact-challenged posts: infowars.com.[4] I asked Li what he thought. He said while the system would be most successful on political stories, it might work. "I think a page from infowars sounds interesting," he said. "Why not give it a shot and let us know what you find out?" To create a fair fight, my editor and I agreed on two rules: I couldn't pick the fake news on my own, and I couldn't test the AI until after I had completed my own fact-check. A longtime fact-checker at *Popular Science* pulled seven spurious science stories from Infowars, from which my editor and I agreed on one with a politicized topic: climate change.

Because Li hadn't had the budget to update ClaimBuster's truth database since late 2016, we chose a piece published before then: "Climate Blockbuster: New NASA Data Shows Polar Ice Has Not Receded Since 1979," from May 2015. Climate-change deniers and fake-news writers often misrepresent real research to bolster their claims. In checking the report, I relied on facts available only in that period.

To keep it short, we used the first 300 words of the Infowars account.[5] For the human portion of the experiment, I checked the selection as I would any article: line by line. I identified fact-based statements—essentially every sentence—and searched for supporting or contradictory evidence from primary sources, such as climate scientists and academic journals. I also followed links in the Infowars story to assess their quality and to see whether they supported the arguments.

Take, for example, the story's first sentence: "NASA has updated its data from satellite readings, revealing that the planet's polar ice caps have not retreated significantly since 1979, when measurements began." Online, the words "data from satellite readings" had a hyperlink. To take a look at the data the story referenced, I clicked the link, which led to a defunct University of Illinois website, Cryosphere Today. Dead end. I emailed the school. The head of the university's Department of Atmospheric Sciences gave me the email address for a researcher who had worked on the site: John Walsh, now chief scientist for the International Arctic Research Center in Alaska, whom I later interviewed by phone. Walsh told me that the "data from satellite readings" wasn't directly from NASA. Rather, the National Snow and Ice Data Center in Boulder, Colorado, had cleaned up raw NASA satellite data for Arctic sea ice. From there, the University of Illinois analyzed and published it. When I asked Walsh whether that data had revealed that the polar ice caps hadn't retreated much since 1979, as Infowars claimed, he said: "I can't reconcile that statement with what the website used to show."

In addition to talking to Walsh, I used Google Scholar to find relevant scientific literature and landed on a comprehensive paper on global sea-ice trends in the peer-reviewed *Journal of Climate* published by the American Meteorological Society and authored by Claire Parkinson, a senior climate scientist at the NASA Goddard Space Flight Center. I interviewed her too. She walked me through how her research compared with the claims in the Infowars story, showing where the latter distorted the data. While it's true that global sea-ice data collection started in 1979, around when the relevant satellites launched, over time the measurements show a general global trend toward retreat, Parkinson said. The Infowars story also conflated data for Arctic and Antarctic sea ice; although the size of polar sea ice varies from year to year, Arctic sea ice has shown a consistent trend toward shrinking that outpaces the Antarctic's trend toward growth, bringing the global totals down significantly. The Infowars author, Steve Watson, conflates Arctic, Antarctic, global, yearly, and average data throughout the article, and may have cherry-picked data from an Antarctic boom year to swell his claim.

In other cases, the Infowars piece linked to poor sources—and misquoted them. 15

Take, for example, a sentence that claims Al Gore warned that the Arctic ice cap might disappear by 2014. The sentence linked to a *Daily Mail* article—not a primary source—that included a quote allegedly from Gore's 2007 Nobel Prize lecture. But when I read the speech transcript and watched the video on the Nobel Prize website, I found that the newspaper had heavily edited the quote, cutting out caveats and context. As for the rest of the Infowars story, I followed the same process. All but two sentences were wrong or misleading. (An Infowars spokesman said the author declined to comment.)

With my own work done, I was curious to see how ClaimBuster would perform. The site requires two steps to do a fact-check. In the first, I copied and pasted the 300-word excerpt into a box labeled "Enter Your Own Text,"[6] to identify factual claims made in the copy. Within one second, the AI scored each line on a scale of zero to one; the higher the number, the more likely it contains a claim. The scores ranged from 0.16 to 0.78. Li suggested 0.4 as threshold for a claim worth further inspection. The AI scored twelve out of sixteen sentences at or above that mark.

In total, there were eleven check-worthy claims among twelve sentences, all of which I had also identified. But ClaimBuster missed four. For instance, it gave a low score of 0.16 to a sentence that said climate change "is thought to be due to a combination of natural and, to a much lesser extent, human influence." This sentence is indeed a claim—a false one. Scientific consensus holds that humans are primarily to blame for recent climate change. False negatives like this, which rate a sentence as not worth checking even when it is, could lead a reader to be duped by a lie.

How could ClaimBuster miss this statement when so much has been written about it in the media and academic journals? Li said his AI likely didn't catch it because the language is vague. "It doesn't mention any specific people or groups," he says. Because the sentence had no hard numbers and cited no identifiable people or institutions, there was "nothing to quantify." Only a human brain can spot the claim without obvious footholds.

Next up, I fed each of the eleven identified claims into a second window, which checks against the system's truth database. In an ideal case, the machine would match the claim to an existing fact-check and flag it as 20 true or false. In reality, it spit out information that was for the most part, irrelevant.

Take the article's first sentence, about the retreat of the polar ice caps. ClaimBuster compared the string of words to all sentences in its database. It searched for matches and synonyms or semantic similarities. Then it ranked hits. The best match came from a PolitiFact story—but the topic concerned nuclear negotiations between the U.S. and Iran, not sea ice or climate change. Li said the system was probably latching onto similar words that don't have much to do with the topic. Both sentences, for example, contain the words "since," "has," "not," as well as similar words such as "updated" and "advanced." This gets at a basic problem: The program doesn't yet weigh more-important words over nonspecific words. For example, it couldn't tell that the Iran story was irrelevant.

When I tried the sentence about Al Gore, the top hit was more promising: Another link from PolitiFact matched to a sentence in a story that read: "Scientists project that the Arctic will be ice-free in the summer of 2013." Here, the match was more obvious; the sentences shared words, including "Arctic," and synonyms such as "disappear" and "ice-free." But when I dug further, it turned out the PolitiFact story was about a 2009 Huffington Post op-ed by then-senator John Kerry, rather than Al Gore in a 2007 Nobel lecture. When I tested the remaining claims in the story, I faced similar problems.

When I reported these result to Li, he wasn't surprised. The problem was that ClaimBuster's truth database didn't contain a report on this specific piece of fake news, or anything similar. Remember, it's made up of work from human fact-checkers at places including PolitiFact and *The Washington Post*. Because the system relies so heavily on information supplied by people, he said, the results were "just another point of evidence that human fact-checkers aren't enough."

That doesn't mean AI fact-checking is all bad. On the plus side. ClaimBuster is way faster than I can ever be. I spent six hours on my fact-check. By comparison, the AI took about 11 minutes. Also consider that I knock off at the end of the day. An AI doesn't sleep. "It's like a tireless intern who will sit watching TV for 24 hours and have a good eye for what a factual claim is," Adair says. As Li's team tests new AI to improve claim scoring and fact-checking, ClaimBuster is bound to improve, as should others. Adair's cooperative is also using ClaimBuster to scan the claims of pundits and politicians on cable TV, highlighting the most check-worthy utterings and emailing them to human fact-checkers to confirm.

Fake news can be difficult to spot, especially when it is marketed alongside, or in the same ways, as credible news.
Michael Brochstein/Sipa via AP Images

25 The trick will be getting the accuracy to match that efficiency. After all, we're in our current predicament, at least in part, because of algorithms. As of late 2017, Google and Facebook had 1.17 billion and 2.07 billion users, respectively. That enormous audience gives fake-news makers and propagandists incentive to game the algorithms to spread their material—it might be possible similarly manipulate an automated fact-checker. And Big Tech's recent attempts to fix their AI haven't gone very well. For example, in October 2017, after a mass shooting in Las Vegas left 851 injured and 58 dead, users from the message board 4chan were able to promote a fake story misidentifying the shooter on Facebook. And last fall, Google AdWords placed fake-news headlines on both PolitiFact and Snopes.

Even if there were an AI fact-checker that's immune to errors and gaming, there would be a larger issue with ClaimBuster and projects like it—and with fake news in general. Political operatives and partisan readers often don't care if an article is intentionally wrong. As long as it supports their agenda—or just makes them snicker—they'll share it. According to the 2017 Princeton, Dartmouth, and Exeter study, people who consumed fake news also consumed so-called hard news—and politically knowledgeable consumers were actually more likely to look at the fake stuff. In other words, it's not like readers don't know the difference. The media should not underestimate their desire to click on such catnip.

One last wrinkle. As companies roll out an army of AI fact-checkers, partisan readers on both sides might view them as just another mode of spin. President Donald Trump has called trusted legacy news outfits such as the *New York Times* and CNN "fake news." Infowars, a site he admires, maintains its own list of fake-news sources, which includes *The Washington Post*. Infowars has also likened the work of fact-checking sites like Snopes and PolitiFact to censorship.

Still, AI fact-checkers might be our best ally in thwarting fake news. There's a lot of digital foolery to track. One startup, Veracity.ai—backed by the Knight Prototype Fund and aimed at helping the ad industry identify fake news that might live next to online ads—recently identified 1,200 phony-news websites and some 400,000 individual fake posts, a number the company expects to grow. It's so fast and cheap to tell a lie, and it's so expensive and time-sucking for humans to correct it. And we could never rely on readers for click-through fact-checking. We'll still need journalists to employ the AI fact-checkers to scour the internet for deception, and to provide fodder for the truth databases.

I asked Li whether my one fact-checked story might have an impact, if it would even make its way into the ClaimBuster truth database. "A perfect automatic tool would capture your data and make it part of the repository," he said.

He added, "Of course, right now, there is no such tool."

30

Tune Your BS Detector

We can't rely on algorithms to flag every falsehood, and there will never be enough journalists to keep up with the deluge. Here's how you can spot fake news.

Look for Sources

Credibility comes when an author cites or quotes multiple primary sources. These include academics, eyewitnesses, and anyone with firsthand knowledge of a topic. It also includes documents such as peer-reviewed studies or reports. For the people: Dig deep and make sure they are real. For studies: Read them on free academic journal databases.

Check the Byline

Reputable news outlets include a reporter's byline. Look 'em up on social media or their professional websites, and check their credentials. Read their other work, which can give you a sense of how they've covered other topics. While some publications don't credit writers (*The Economist*), a wild account with no author listed is a hint that it isn't credible.

Look for Corrections

Everyone makes mistakes, and journalists are supposed to acknowledge theirs by running corrections. Fake-news writers and propagandists don't update their work even in the face of legitimate criticism; they aren't trying to verify the record but manipulate it. If you can't find any corrections by the news outlet or in the author's other work, beware.

Look at Outlets Covering It

If an outrageous claim is true—particularly one with political implications—you can bet major media will be all over it. Their job is to break news and follow up on other outlets that do so. If a website reports a jaw-dropping revelation and no journalists pick it up, there is likely a good reason: Maybe they vetted the story and it's not true.

Be a Photo Sleuth

Images can be manipulated. You can chase down the origin of many photos using Google's reverse image search (reverse. photos). This will show the original publication and data for photographs, screenshots, memes, and more—proving once and for all that the latest hurricane did not, in fact, allow sharks to swim down flooded highways.

Be a Media Omnivore

Whatever your politics, it's good to burst your filter bubble. Left-leaning? See what *National Review* is serving up. Right-leaning? Check out *Mother Jones*. You might not agree with all the articles, but you'll get a better sense of the range of worldviews—which will also help you pinpoint the difference between a reported story and propaganda.

Notes

1. "Fake news" is an embattled term. It is used to describe news that is intentionally meant to mislead—for political or economic gain—based on false, misinterpreted, or manipulated facts. But partisans also use it to smear reputable legacy media outlets. Here, we're using the former definition.

2. The book is part of a family of writing guides from the University of Chicago Press. And yes, the facts in it are valid beyond Chicago.
3. A *Popular Science* fact-checker spent fifteen hours verifying the pages you're reading and caught thirty-four errors before we went to press.
4. Infowars is a media empire and clearinghouse for conspiracies—from the federal government controlling weather to the idea that Glenn Beck is a CIA operative.
5. We made sure that the rest of the story did not provide evidence or context that would affect our fact-check.
6. http://idir-server2.uta.edu/claimbuster/

Understanding the Text

1. Why does Borel believe that truth is "complex and squishy" (par. 5)?
2. "Human fact-checkers aren't enough" according to the article (par. 23). Why not?
3. Why might someone not care if "an article is intentionally wrong" (par. 26)?

Reflection and Response

4. Borel references a report from the Pew Research Center that found that the majority of Americans (two-thirds) get their news from social media (par. 2). Where do you get your news from? Keep a news log for one week where you log which news you consume, from where, and how often. What does your log tell you about your habits?
5. In the article, Borel recounts the extensive work that went into verifying the information "reported" in an article from the website *Infowars.com*. Borel is aware this is a website that regularly promotes conspiracy theories, but someone lacking that context may take the article at face value. Given this example, what is your own process, if any, for fact-checking stories you come across online, especially on websites you are unfamiliar with?
6. Due to the innately subjective nature of human opinion, Borel claims that "AI fact-checkers might be our best ally in in thwarting fake news" (par. 28). Do you agree or disagree?

Making Connections

7. Although ClaimBuster utilizes AI and advanced algorithms, it was nevertheless fallible: Borel caught multiple errors as she duplicated the work. Identify two or three other forms of technology or software (for example, spell check and autocorrect) that humans rely on to complete daily tasks. Then, list some of the risks that may result if humans do not double-check the results.
8. Borel mentions that she hosts a podcast called *Methods* that "explores how journalists, scientists, and other professional truth-finders know what they know" (par. 5). In addition to listening to *Methods*, watch "How to Seek

Truth in the Era of Fake News," the TED talk interview with CNN veteran journalist Christiane Amanpour. Then, visit two fact-checking websites like Snopes.com or Factcheck.org. What do these sources teach you about the complexity of truth? Why might it be difficult for someone to determine what is truthful, even with fact-checking?

9. Locate Tom Hale's article "Marijuana Contains 'Alien DNA' from Outside of Our Solar System, NASA Confirms," posted in 2016 on *IFLScience*. After reading the article itself, what is the purpose of it? What connections do you see between this article and Borel's?

Sentence Guides for Academic Writers

Being a college student means being a college writer. No matter what field you are studying, your instructors will ask you to make sense of what you are learning through writing. When you work on writing assignments in college, you are, in most cases, being asked to write for an academic audience.

Writing academically means thinking academically — asking a lot of questions, digging into the ideas of others, and entering into scholarly debates and academic conversations. As a college writer, you will be asked to read different kinds of texts; understand and evaluate authors' ideas, arguments, and methods; and contribute your own ideas. In this way, you present yourself as a participant in an academic conversation.

What does it mean to be part of an *academic conversation*? Well, think of it this way: You and your friends may have an ongoing debate about the best film trilogy of all time. During your conversations with one another, you analyze the details of the films, introduce points you want your friends to consider, listen to their ideas, and perhaps cite what the critics have said about a particular trilogy. This kind of conversation is not unlike what happens among scholars in academic writing — except they could be debating the best public policy for a social problem or the most promising new theory in treating disease.

If you are uncertain about what academic writing *sounds like* or if you're not sure you're any good at it, this booklet offers guidance for you at the sentence level. It helps answer questions such as these:

> How can I present the ideas of others in a way that demonstrates my understanding of the debate?
>
> How can I agree with someone, but add a new idea?
>
> How can I disagree with a scholar without seeming, well, rude?
>
> How can I make clear in my writing which ideas are mine and which ideas are someone else's?

The following sections offer sentence guides for you to use and adapt to your own writing situations. As in all writing that you do, you will have to think about your purpose (reason for writing) and your audience (readers) before knowing which guides will be most appropriate for a particular piece of writing or for a certain part of your essay.

The guides are organized to help you present background information, the views and claims of others, and your own views and claims — all in the context of your purpose and audience.

Academic Writers Present Information and Others' Views

When you write in academic situations, you may be asked to spend some time giving background information for or setting a context for your main idea or argument. This often requires you to present or summarize what is known or what has already been said in relation to the question you are asking in your writing.

SG1 Presenting What Is Known or Assumed

When you write, you will find that you occasionally need to present something that is known, such as a specific fact or a statistic. The following structures are useful when you are providing background information.

As we know from history, ________.

X has shown that ________.

Research by X and Y suggests that ________.

According to X, ________ percent of ________ are/favor ________.

In other situations, you may have the need to present information that is assumed or that is conventional wisdom.

People often believe that ________.

Conventional wisdom leads us to believe ________.

Many Americans share the idea that ________.

________ is a widely held belief.

In order to challenge an assumption or a widely held belief, you have to acknowledge it first. Doing so lets your readers believe that you are placing your ideas in an appropriate context.

Although many people are led to believe X, there is significant benefit to considering the merits of Y.

College students tend to believe that ________ when, in fact, the opposite is much more likely the case.

SG2 Presenting Others' Views

As a writer, you build your own *ethos*, or credibility, by being able to fairly and accurately represent the views of others. As an academic writer, you will be expected to demonstrate your understanding of a text by summarizing the views or arguments of its author(s). To do so, you will use language such as the following.

X argues that ________.

X emphasizes the need for ________.

In this important article, X and Y claim ________.

X endorses ________ because ________.

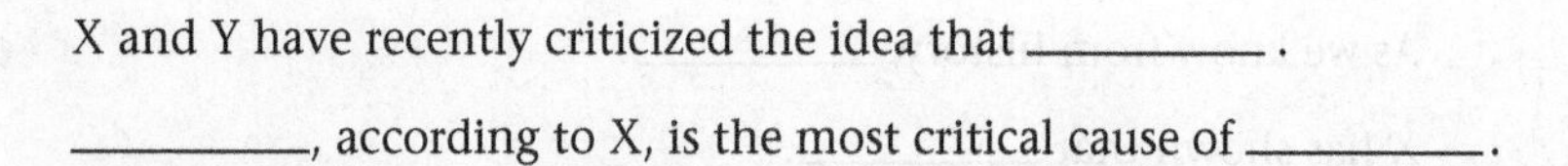

X and Y have recently criticized the idea that ________.

________, according to X, is the most critical cause of ________.

Although you will create your own variations of these sentences as you draft and revise, the guides can be useful tools for thinking through how best to present another writer's claim or finding clearly and concisely.

SG3 Presenting Direct Quotations

When the exact words of a source are important for accuracy, authority, emphasis, or flavor, you will want to use a direct quotation. Ordinarily, you will present direct quotations with language of your own that suggests how you are using the source.

X characterizes the problem this way: ". . ."

According to X, ________ is defined as ". . ."

". . . ," explains X.

X argues strongly in favor of the policy, pointing out that ". . ."

Note: You will generally cite direct quotations according to the documentation style your readers expect. MLA style, often used in English and in other humanities courses, recommends using the author name paired with a page number, if there is one. APA style, used in most social sciences, requires the year of publication generally after the mention of the source, with page numbers after the quoted material. In *Chicago* style, used in history and in some humanities courses, writers use superscript numbers (like this[6]) to refer readers to footnotes or endnotes. In-text citations, like the ones shown below, refer readers to entries in the works cited or reference list.

MLA	Lazarín argues that our overreliance on testing in K-12 schools "does not put students first" (20).
APA	Lazarín (2014) argues that our overreliance on testing in K-12 schools "does not put students first." (p. 20)
Chicago	Lazarín argues that our overreliance on testing in K-12 schools "does not put students first."[6]

Many writers use direct quotations to advance an argument of their own:

Standardized testing makes it easier for administrators to measure student performance, but it may not be the best way to measure it. Too much testing wears students out and communicates the idea that recall is the most important skill we want them to develop. Even education policy advisor Melissa Lazarín argues that our overreliance on testing in K-12 schools "does not put students first" (20).

Student writer's idea

Source's idea

SG4 Presenting Alternative Views

Most debates, whether they are scholarly or popular, are complex—often with more than two sides to an issue. Sometimes you will have to synthesize the views of multiple participants in the debate before you introduce your own ideas.

> On the one hand, X reports that ________, but on the other hand, Y insists that ________.
>
> Even though X endorses the policy, Y refers to it as ". . ."
>
> X, however, isn't convinced and instead argues ________.
>
> X and Y have supported the theory in the past, but new research by Z suggests that ________.

Academic Writers Present Their Own Views

When you write for an academic audience, you will indeed have to demonstrate that you are familiar with the views of others who are asking the same kinds of questions as you are. Much writing that is done for academic purposes asks you to put your arguments in the context of existing arguments—in a way asking you to connect the known to the new.

When you are asked to write a summary or an informative text, your own views and arguments are generally not called for. However, much of the writing you will be assigned to do in college asks you to take a persuasive stance and present a reasoned argument—at times in response to a single text, and at other times in response to multiple texts.

SG5 Presenting Your Own Views: Agreement and Extension

Sometimes you agree with the author of a source.

X's argument is convincing because ________.

Because X's approach is so ________, it is the best way to ________.

X makes an important point when she says ________.

Other times you find you agree with the author of a source, but you want to extend the point or go a bit deeper in your own investigation. In a way, you acknowledge the source for getting you so far in the conversation, but then you move the conversation along with a related comment or finding.

X's proposal for ________ is indeed worth considering. Going one step further, ________.

X makes the claim that ________. By extension, isn't it also true, then, that ________?

________ has been adequately explained by X. Now, let's move beyond that idea and ask whether ________.

SG6 Presenting Your Own Views: Queries and Skepticism

You may be intimidated when you're asked to talk back to a source, especially if the source is a well-known scholar or expert or even just a frequent voice in a particular debate. College-level writing asks you to be skeptical, however, and approach academic questions with the mind of an investigator. It is OK to doubt, to question, to challenge—because the end result is often new knowledge or new understanding about a subject.

Couldn't it also be argued that ________?

But is everyone willing to agree that this is the case?

While X insists that ________ is so, he is perhaps asking the wrong question to begin with.

The claims that X and Y have made, while intelligent and well-meaning, leave many unconvinced because they have failed to consider ________.

A Note about Using First Person "I"

Some disciplines look favorably upon the use of the first person "I" in academic writing. Others do not and instead stick to using third person. If you are given a writing assignment for a class, you are better off asking your instructor what he or she prefers or reading through any samples given than *guessing* what might be expected.

First person (*I, me, my, we, us, our*)

I question Heddinger's methods and small sample size.

Harnessing children's technology obsession in the classroom is, I believe, the key to improving learning.

Lanza's interpretation focuses on circle imagery as symbolic of the family; my analysis leads me in a different direction entirely.

We would, in fact, benefit from looser laws about farming on our personal property.

Third person (names and other nouns)

Heddinger's methods and small sample size are questionable.

Harnessing children's technology obsession in the classroom is the key to improving learning.

Lanza's interpretation focuses on circle imagery as symbolic of the family; other readers' analyses may point in a different direction entirely.

Many Americans would, in fact, benefit from looser laws about farming on personal property.

You may feel as if not being able to use "I" in an essay in which you present your ideas about a topic is unfair or will lead to weaker statements. Know that you can make a strong argument even if you write in the third person. Third person writing allows you to sound more assertive, credible, and academic.

Presenting Your Own Views: Disagreement or Correction

You may find that at times the only response you have to a text or to an author is complete disagreement.

X's claims about ________ are completely misguided.

X presents a long metaphor comparing ________ to ________; in the end, the comparison is unconvincing because ________.

It can be tempting to disregard a source completely if you detect a piece of information that strikes you as false or that you know to be untrue.

Although X reports that ________, recent studies indicate that is not the case.

While X and Y insist that is ________ so, an examination of their figures shows that they have made an important miscalculation.

SG8 Presenting and Countering Objections to Your Argument

Effective college writers know that their arguments are stronger when they can anticipate objections that others might make.

Some will object to this proposal on the grounds that ________.

Not everyone will embrace ________; they may argue instead that ________.

Countering, or responding to, opposing voices fairly and respectfully strengthens your writing and your *ethos*, or credibility.

X and Y might contend that this interpretation is faulty; however, ________.

Most ________ believe that there is too much risk in this approach. But what they have failed to take into consideration is ________.

Academic Writers Persuade by Putting It All Together

Readers of academic writing often want to know what's at stake in a particular debate or text. Aside from crafting individual sentences, you must, of course, keep the bigger picture in mind as you attempt to persuade, inform, evaluate, or review.

SG9 Presenting Stakeholders

When you write, you may be doing so as a member of a group affected by the research conversation you have entered. For example, you may be among the thousands of students in your state whose level of debt may change as a result of new laws about financing a college education. In this case, you are a *stakeholder* in the matter. In other words, you have an interest in the matter as a person who could be impacted by the outcome of a decision. On the other hand, you may be writing as an investigator of a topic that interests you but that you aren't directly connected with. You may be persuading your audience on behalf of a group of interested stakeholders—a group of which you yourself are not a member.

You can give your writing some teeth if you make it clear who is being affected by the discussion of the issue and the decisions that have or will be made about the issue. The groups of stakeholders are highlighted in the following sentences.

> Viewers of Kurosawa's films may not agree with X that ________.
>
> The research will come as a surprise to parents of children with Type 1 diabetes.
>
> X's claims have the power to offend potentially every low-wage earner in the state.
>
> Marathoners might want to reconsider their training regimen if stories such as those told by X and Y are validated by the medical community.

SG10 Presenting the "So What"

For readers to be motivated to read your writing, they have to feel as if you're either addressing something that matters to them or addressing something that matters very much to you or that should matter to us all. Good academic writing often hooks readers with a sense of urgency—a serious response to a reader's "So what?"

> Having a frank discussion about ________ now will put us in a far better position to deal with ________ in the future. If we are unwilling or unable to do so, we risk ________.
>
> Such a breakthrough will affect ________ in three significant ways.

It is easy to believe that the stakes aren't high enough to be alarming; in fact, ________ will be affected by ________.

Widespread disapproval of and censorship of such fiction/films/art will mean ________ for us in the future. Culture should represent ________.

________ could bring about unprecedented opportunities for ________ to participate in ________, something never seen before.

New experimentation in ________ could allow scientists to investigate ________ in ways they couldn't have imagined ________ years ago.

SG11 Presenting the Players and Positions in a Debate

Some disciplines ask writers to compose a review of the literature as a part of a larger project—or sometimes as a freestanding assignment. In a review of the literature, the writer sets forth a research question, summarizes the key sources that have addressed the question, puts the current research in the context of other voices in the research conversation, and identifies any gaps in the research.

Writing that presents a debate, its players, and their positions can often be lengthy. What follows, however, can give you the sense of the flow of ideas and turns in such a piece of writing.

________ affects more than 30% of children in America, and signs point to a worsening situation in years to come because of A, B, and C. Solutions to the problem have eluded even the sharpest policy minds and brightest researchers. In an important 2003 study, W found that ________, which pointed to more problems than solutions. [. . .] Research by X and Y made strides in our understanding of ________ but still didn't offer specific strategies for children and families struggling to ________. [. . .] When Z rejected both the methods and the findings of X and Y, arguing that ________, policy makers and health-care experts were optimistic. [. . .] Too much discussion of ________, however, and too little discussion of ________, may lead us to solutions that are ultimately too expensive to sustain.

Student writer states the problem.

Student writer summarizes the views of others on the topic.

Student writer presents her view in the context of current research.

Appendix: Verbs Matter

Using a variety of verbs in your sentences can add strength and clarity as you present others' views and your own views.

When you want to present a view fairly neutrally

acknowledges
adds
admits
comments
contends
notes
observes
points out
reports
suggest
writes

X points out that the plan had unintended outcomes.

When you want to present a stronger view

argues
asserts
declares
emphasizes
insists

Y argues in favor of a ban on ________; but Z insists the plan is misguided.

When you want to show agreement

agrees
confirms
endorses

An endorsement of X's position is smart for a number of reasons.

When you want to show contrast or disagreement

compares
denies
disputes
refutes
rejects

The town must come together and reject X's claims that ________ is in the best interest of the citizens.

When you want to anticipate an objection

admits
acknowledges
concedes

Y admits that closer study of ________, with a much larger sample, is necessary for ________.

Acknowledgments

Steven Aftergood, "Privacy and the Imperative of Open Government" from *Privacy in the Modern Age: The Search for Solutions.* Copyright © 2015 by Marc Rotenberg, Julia Horwitz, and Jeramie Scott. Reprinted by permission of The New Press. www .thenewpress.com

Liz Alderman, "Sweden's Push to Get Rid of Cash Has Some Saying, 'Not So Fast'," *The New York Times,* November 21, 2018. © 2018 The New York Times. All rights reserved. Used under license.

Chris Anderson, "Drones go to Work: The Disruptive Economics of unmanned vehicles are taking hold. Here's how to think about the drone economy and your place in it," *Harvard Business Review,* May 2017. Used with permission.

Emily Anthes, "Animals Bow to Their Mechanical Overlords," originally appeared in *Nautilus,* February 13, 2014. Copyright © 2014 by Emily Anthes. Used with permission.

Ole Bjerg, "How is Bitcoin Money?" *Theory, Culture & Society* 33(1), pp. 53–72. Copyright © 2015 by the Author. Reprinted by permission of SAGE Publications, Ltd. http:// www.sagepub.co.uk. This material is the exclusive property of the SAGE Publishing and is protected by copyright and other intellectual property laws. User may not modify, publish, transmit, participate in the transfer or sale of, reproduce, create derivative works (including course packs) from, distribute, perform, display, or in any way exploit any of the content of the file(s) in whole or in part. Permission may be sought for further use from Publications Ltd., Rights & Permissions Department, 1, Oliver's Yard, 55, City Road, London EC1Y 1SP, Email: permissions@sagepub .co.uk. By accessing the file(s), the User acknowledges and agrees to these terms.

Brooke Borel, "Can artificial intelligence help solve the internet's misinformation problem?" *Popular Science,* Spring 2018. Reprinted with the permission of the author.

Matt Britton, "The Peer-to-Peer Economy" from *YouthNation: Building Remarkable Brands in a Youth-Driven Culture* by Matt Britton. Copyright © 2015 Wiley. Reproduced with permission of John Wiley & Sons, Inc.

Ronald Brownstein, "How Has Technology Changed the Concept of Community?" *The Atlantic,* October 10, 2015. © 2015 The Atlantic Media Co., as first published in *The Atlantic Magazine.* All rights reserved. Distributed by Tribune Content Agency, LLC.

Jon Cohen, "The Horror Story That Haunts Science: 200 years later *Frankenstein* still shocks and inspires," *Science,* January 2018. Republished with permission of the American Association for the Advancement of Science; permission conveyed through Copyright Clearance Center, Inc.

Republished with permission of The Economist, from "A-I Spy," *The Economist,* March 31st–April 6th, 2018 (section: Future Workplaces: Smile, You're On Camera); permission conveyed through Copyright Clearance Center, Inc.

William Eyre, "Surveillance Today" from *The Real ID Act: Privacy and Government Surveillance* by William Eyre. Publisher: LFB Scholarly Publishing LLC (March 15, 2011). Reprinted with permission.

Lisa Fickenscher, "Is Amazon Good for America? Lost jobs, Shuttered Shops, and Ailing $tates (states) all part of 'success' story," *New York Post,* April 25 2017. © 2017 New York Post. All rights reserved. Used by permission and protected by the Copyright Laws of the United States. The printing, copying, redistribution, or retransmission of this Content without express written permission is prohibited.

Martin Ford, "The Healthcare Challenge" from *Rise of the Robots: Technology and the Threat of a Jobless Future* by Martin Ford, copyright © 2015. Reprinted by permission of Basic Books, an imprint of Hachette Book Group, Inc.

Malcolm Gladwell, "Small Change: Why the Revolution will not be Tweeted," *The New Yorker,* October 4, 2010. Reprinted with permission of the author.

Clive Hamilton, An excerpt from Ch. 1 of *Earthmasters: The Dawn of the Age of Climate Engineering* by Clive Hamilton. Publisher: Yale University Press (March 25, 2014). Reprinted with permission of the publisher.

Meagan Johnson, "Stop Talking About Work/Life Balance! TEQ and the Millennial Generation," *Workforce Solutions Review,* an IHRIM publication, March 2015. www.ihrimpublications.com.

Rohit Karnik and Robert S. Langer, "Rebuilding Ourselves: Ushering in an Age of Synthetic Organs and Targeted Medicine," *Mechanical Engineering,* February 2013. pp. 35–39. Reprinted with permission.

Natalie Kitroeff, "Robots Could Replace 1.7 Million American Truckers in the Next Decade," *LA Times,* September 25, 2016. Copyright © 2016 by Los Angeles Times. Used with permission.

Sheldon Krimsky, "Creating Good from Immoral Acts" from *Stem Cell Dialogues* by Sheldon Krimsky. Copyright © June 23, 2015 Columbia University Press. Reprinted with permission of Columbia University Press.

Scott Lucas, "Why Cambridge Analytica Matters," *EA Worldview,* April 3, 2018. Copyright © 2013–2018 Scott Lucas and EaWorldView.com. All Rights Reserved. Used with permission.

The Media Insight Project, "How Americans Get Their News" from *The Personal News Cycle.* Copyright © 2014 by The Media Insight Project. Reprinted with permission of the Media Insight Project.

Leo Mirani, "Sorry, Malcolm Gladwell, the Revolution May Well Be Tweeted," *The Guardian,* October 2, 2010. Copyright Guardian News & Media Ltd 2018. Used with permission.

Paul Mozur, Mark Scott, and Mike Isaac, "Facebook Faces a New World as Officials Rein In a Wild Web," *The New York Times,* September 17, 2017. © 2017 The New York Times. All rights reserved. Used by permission and protected by the Copyright Laws of the United States. The printing, copying, redistribution, or retransmission of this Content without express written permission is prohibited. www.nytimes.com

Sarah E. Myhre, "Treading the Fine Line Between Climate Talk and Alarmism," *Live Science,* June 23, 2017. Reprinted with the permission of EnVeritas Group.

Alice Park, "The CRISPR Pioneers," *TIME.com,* 2016. © 2016 Time Inc. Used under license. TIME.com and Time Inc. are not affiliated with, and do not endorse products or services of, Licensee.

Nicole Perlroth, "Governments Turn to Commercial Spyware to Intimidate Dissidents," *The New York Times,* May 30, 2016. © 2016 The New York Times. All rights reserved. Used by permission and protected by the Copyright Laws of the United States. The printing, copying, redistribution, or retransmission of this Content without express written permission is prohibited. www.nytimes.com

Francesc Torralba Roselló, "The Argument of Future Generations" an extract from his chapter "Foundations of Eco-Ethics: A Global Analysis" in *Environmental Ethics Revisited* edited by Rosamund Thomas. Publisher Ethics International Press Ltd. (February 28, 2014). Reprinted with permission from the publisher.

Hillary Rosner, "All Too Human: If We Could, Would We Want to Live Forever?" *Scientific American* 315(3):70–75. September 2016. Reproduced with permission. Copyright © 2016 Scientific American, a division of Nature America, Inc. All rights reserved.

Sam Ross-Brown, "Net Neutrality and the Fight for Social Justice," *Tikkun,* Volume 30, no. 3. Copyright © 2015, Tikkun Magazine. All rights reserved. Republished by permission of the copyright holder, and the present publisher, Duke University Press. www.dukeupress.edu

Beth Shapiro. Republished with permission of Princeton University Press, excerpt from Ch. 1 of *How to Clone a Mammoth: The Science of De-Extinction* by Beth Shapiro, Princeton University Press, April 2015; permission conveyed through Copyright Clearance Center, Inc.

Gary Shteyngart, excerpt(s) from *Super Sad True Love Story: A Novel* by Gary Shteyngart, copyright © 2010 by Gary Shteyngart. Used by permission of Random House, an imprint and division of Penguin Random House LLC. All rights reserved.

Stuart Sumner, excerpts from the Introduction: "Why All this Fuss About Privacy," "Here's my Cow, Now Where's My Change?" and "Hey I Thought this Lunch was Free!" from *You: For Sale* by Stuart Sumner. Copyright 2015, with permission from Elsevier.

Thomas Sumner, "Quenching Society's Thirst: Desalination may soon turn a corner, from rare to routine," *Science News,* August 9, 2016. Used with permission.

Kurt Vonnegut, "Harrison Bergeron" from *Welcome to the Monkey House* by Kurt Vonnegut, copyright © 1968 by Kurt Vonnegut Jr. Copyright renewed © 1996 by Kurt Vonnegut Jr.. Used by permission of Dell Publishing, an imprint of Random House, a division of Penguin Random House LLC. All rights reserved.

Lori Ann Wagner, "When Your Smart Phone Is Too Smart for Your Own Good: How Social Media Alters Human Relationships," *The Journal of Individual Psychology* Volume 71, Issue 2, Summer 2015, pp. 114–121. Copyright © 2015 by the University of Texas Press. All rights reserved.

Bryan Walsh, "New Natural Selection: How Scientists Are Altering DNA to Genetically Engineer New Forms of Life," *Newsweek,* June 29, 2017. © 2017 Newsweek Media Group. All rights reserved. Used by permission and protected by the Copyright Laws of the United States. The printing, copying, redistribution, or retransmission of this Content without express written permission is prohibited. www.newsweek.com

Selina Wang, "The Future of Farming is Looking Up," *Bloomberg,* September 6, 2017. Used with permission of Bloomberg L.P. Copyright © 2017. All rights reserved.

Rob Wile and Pascal Lee, "Here's How Much it Would Cost to Travel to Mars," *MONEY.com,* May 8, 2017. © 2017 Time Inc. Used under license. MONEY.com and Time Inc. are not affiliated with, and do not endorse products or services of, Licensee.

Index of Authors and Titles